A Compendious System of Astronomy

In a Course of Familiar Lectures

M ARGARET B RYAN

CAMBRIDGE
UNIVERSITY PRESS

CAMBRIDGE UNIVERSITY PRESS

Cambridge, New York, Melbourne, Madrid, Cape Town,
Singapore, São Paolo, Delhi, Mexico City

Published in the United States of America by Cambridge University Press, New York

www.cambridge.org
Information on this title: www.cambridge.org/9781108050333

© in this compilation Cambridge University Press 2012

This edition first published 1797
This digitally printed version 2012

ISBN 978-1-108-05033-3 Paperback

This book reproduces the text of the original edition. The content and language reflect
the beliefs, practices and terminology of their time, and have not been updated.

Cambridge University Press wishes to make clear that the book, unless originally published
by Cambridge, is not being republished by, in association or collaboration with, or
with the endorsement or approval of, the original publisher or its successors in title.

Engraved by W.ᵗ Nutter from a Miniature of the same size painted by Samˡ Shelley.

MᴿˢBRYAN and CHILDREN.

A

COMPENDIOUS SYSTEM

OF

ASTRONOMY,

IN A

COURSE OF FAMILIAR LECTURES;

IN WHICH THE PRINCIPLES OF THAT SCIENCE ARE CLEARLY
ELUCIDATED, SO AS TO BE INTELLIGIBLE TO THOSE
WHO HAVE NOT STUDIED THE MATHEMATICS.

ALSO

TRIGONOMETRICAL and CELESTIAL PROBLEMS,

WITH

A KEY to the EPHEMERIS,

AND

A VOCABULARY OF THE TERMS OF SCIENCE

USED IN THE LECTURES;

WHICH LATTER ARE EXPLAINED AGREEABLY TO THEIR APPLICATION IN THEM.

By MARGARET BRYAN.

These are thy glorious Works, PARENT of GOOD,.
ALMIGHTY! thine this universal frame,
Thus wond'rous fair; THYSELF how wond'rous then!
Unspeakable! who sit'st above these Heav'ns,
To us invisible, or dimly seen
In these thy lowest works; yet these declare
Thy Goodness beyond thought, and Pow'r Divine.   MILTON.

LONDON.

PRINTED FOR THE AUTHOR,
And fold by LEIGH and SOTHEBY, York Street, Covent Garden;
And G. KEARSLEY, No. 46, Fleet Street.
1797.
[Entered at Stationers Hall.]

DEDICATION.

To my *PUPILS.*

ASTRONOMY being the most important Science I have had the pleasure of introducing to your acquaintance, I think, by publishing my Lectures on that subject, I shall afford you a pleasing retrospect of the sublime ideas it conveys: And as I know the recollection of the friend who delivered them will not be unpleasing to you, agreeably to your wishes, I have prefixed my portrait to these Lectures, together with those of my two dear children; hoping by the former to make a more lasting impression on your minds---as, believe me, to be forgotten by you, would inflict a severe pang

on that heart which feels for you almost parental
tenderness --- the truth of which I trust you cannot
doubt, as your own breasts bear testimony to the
probability, and must naturally infer the certainty,
of this assertion. Could it be possible to receive
the repeated proofs of your affection which I have
received, and not love you? No! Can I ever
forget the gratulations you have annually offered
me, when each countenance, glowing with affection or dissolving in tenderness, according as the
grateful ideas flowed in your breasts, expressed
more than your words? Believe me, at those
moments I have felt the most sensible sympathy
thrill at my heart --- such emanations as I then received, glowed too warmly in my bosom ever to
be extinguished.

How I have returned your affection you best
know. I had but one way of evincing my since-
rity, which was, by an industrious application to
such studies as would most benefit you, in which

employment I was animated by the hope of a reward, and such a reward as deserved all my assiduity to attain---no less than the present and future happiness of my dear Pupils; which, that they may all enjoy, is the ardent wish of their faithful and affectionate

MARGARET BRYAN.

August, 1797.

PREFACE.

I AM well aware, that to write on subjects which have been so extensively considered, and fully delineated, by the ablest Mathematicians, and by Philosophers of the most penetrating genius, will not procure me any honor on the score of originality; yet I trust I shall not incur censure by publishing a Compendium which, according to my ideas, will render subjects, generally thought obscure, clear to the understanding of young people:—If I have failed in the attempt, through the imbecility of my judgment, I hope the motive may be my apology.

I crave the indulgence of a discriminating public to a Work, the result of much thought and close reasoning—and as the ideas formed of subjects confessedly among the loftiest to the human comprehension, must be imperfect; that my remarks on, and manner of digesting those which are the most popular—may meet a candid reception.

I know that I have no claim to the public suffrage, only on account of the clearness of my illustrations, which, as well as the diagrams, are principally original. As to the phraseology, I fear it is too deficient in ornament to procure me any credit; yet I hope the clearness of the elucidations may gloss over the imperfections of the stile in which they are delivered:—Had I copied that of other authors, I might perhaps have rendered these Lectures more pleasing, although less intelligible to my pupils; who, being familiar with my diction, understand my illustrations much better, as I have thence been able to deliver them more naturally and forcibly.

For the minuteness of my descriptions—the inferences I have taken the liberty to make—and the digressions I have indulged in by way of reflection—I hope to be excused by those whose superior learning does not need the first,—whose judgment would lead them to make better than the second, and whose mature reason renders the last unnecessary to them;—begging them to observe that these Lectures were written for my pupils, and not originally designed for public inspection.

The most tender claims upon my exertions, (the nature of which may be understood from the frontispiece) seconded by the encouragement of my friends, having tempted me to risk the decision of the public, with diffidence I submit to its fiat;—not wishing to procure one devoid of judgment,—but such as the light of the understanding, mingling with the feelings of sentiment without obscuring them, may justify.

I have not presumptuously offered opinions, without having previously digested those of the best writers on the subject; or attempted to elucidate without due observation of the principles of this Science. For this mental exertion, I expect some countenance from those whose extensive learning and liberality lead them to judge impartially—for they, rising superior to the false and vulgar prejudices of many, who suppose these subjects too sublime for female introspection, (ascribing to mental powers the feebleness which characterizes the constitution) invalidate the idea, by affording all laudable exertions their avowed patronage—acknowledging truth, although enfeebled by female attire.

At the tribunal of public and just criticism, I bow with humble confidence,—acknowledging its discriminating judgment,—admiring its general impartiality—yet, conscious of my demerits, dreading its award. Therefore, whilst I invoke its pruning aid to lop off all redundancies, I implore its saving influence, to preserve unhurt this first shoot of a plant, which, if by its fostering mandate benevolently introduced into the mild, the invigorating soil of public esteem, may improve;—but should the rude breath of censure blow on its present weak and humble state,—striking at the root of its vitality, it would repress its faint but salutary odour.

Thus far I have expressed my thoughts as nearly as possible to convey my ideas, and with ease to myself; but a harder task remains to be performed—to express my sense of the goodness of those numerous friends who have graced my subscription by their suffrage! If they can conceive the emotions of joy and gratitude at this moment vibrating in unison in my bosom, they will know them to be such as humanity glories in—all glowing and felicitating, yet tributary and softened to the most grateful sense of obligation. To the sensibility of those who have on this occasion manifested their sympathy and friendship for me, I must appeal, to fill up my deficiency of expression—words being too feeble conveyancers for the mind's energies.

One gentleman in particular claims my warmest emotions of gratitude, as to him I am indebted for all the advantages procured me by this publication; by his having kindly undertaken to peruse the MS. and after perusal, affording it his avowed sanction and patronage; without which distinction, I could not have ventured to offer a work of Science to the public, from the presumption it would have implied, as well as from the apprehension that the imperfection of my judgment might have misled me. This friend, I am proud to acknowledge, in the learned Dr. Charles Hutton, F. R. S. of London and Edinburgh, and of the Philosophical Societies of Haerlem and America, and Professor of Mathematics in the Royal Military Academy, Woolwich. My introduction to this gentleman I esteem one of the happiest circumstances of my life—an epoch I shall ever recur to with extreme satisfaction. Although our acquaintance has been very short, yet have his active services, benevolently exerted to procure me extensive benefit and suffrage, afforded me never-ceasing cause of congratulation on the acquisition of so valuable a friend—whose kindness is the more distinguishing, as I had no claims of friendship, or even personal acquaintance, to intitle me to his notice, previous to the submitting my MS. to his perusal. How sensibly do I feel my obligations to him,—they are such as bind me his debtor for ever.

I hope the candid public will not accuse me of vanity, in publishing a paragraph from the letter my worthy friend sent me on returning my MS.* as I am induced to do so by the advantage it may procure me.

Having premised all I think necessary to the justification of this undertaking—to excuse its faults, and to enforce its precepts; with extreme sensibility, I subscribe myself my Friends and Patrons most grateful, and the Public's

Most obedient humble servant,
MARGARET BRYAN.

* (COPY.)

Woolwich, Jan. 6, 1797.

" *Madam,*

I herewith return the ingenious MS. of Astronomical Lectures you favored me with the sight of, which I have read over with great pleasure; and the more so, to find that even the learned and more difficult Sciences are thus beginning to be successfully cultivated by the extraordinary and elegant talents of the female writers of the present day.

Should you, Madam, give to your friends and to the public to benefit by the publication of these your learned and useful labours, I beg to have the honor of being considered one of the encouragers of so useful a work;

And am, with great respect,

Madam,

Your most obedient, and most humble servant,

Charles Hutton."

SUBSCRIBERS.

A.

HIS Grace the Lord Archbishop of Canterbury
The Right Honorable Lady Arden
Sir John St. Aubyn, Bart. *Stratford Place*
Mr. Dudley Adams, *Fleet Street*
Mrs. Adams, *Clapham*
Alexander Alcock, Efq. *Waterford*
William-Congreve Alcock, Efq. *Waterford*
Mr. Allen, *Errol, Scotland*
Mr. Allen, *Inchmarlin, Scotland*
John Anderson, M. D. *Margate*
Mrs. Angier, *Canterbury*
James Anderson, Esq. *Margate*
Mr. Arthur, *London*
Colonel Auriol, Oxford Light Dragoons
Mrs. Auriol
Alexander Aubert, Esq. F. A. S. F. R. S. *Highbury*
Nicholas Austen, Esq. *Margate*

B.

Right Honorable Lord Americus Beauclerk
Lady Brooke, *Margate*
Honorable Miss Burton, *Ramsgate*
General Maxwell Brown
John Baker, Esq. *Margate*
Mrs. Balderston, *Canterbury*
Mrs. Baker, *Canterbury*
Thomas Barrett, Esq. *Lee Court, Kent*
Thomas Barbut, Esq. *Swaffham, Norfolk*
Mr. W. Balston, *Maidstone*
Major Benezet, *Margate*

Colonel Beaufoy, *Hackney*
R. Becher, Esq. Oxford Light Dragoons
Mrs. C. Beckingham, *Oswalds, Kent*
Reverend Dr. Bell, *Westminster*
Miss Belsey, *Margate*
Miss E. Belsey, *Ditto*
———— Bell, Esq. *Blackheath*
Mr. Benson, *Margate*
Mrs. Benson, *Ditto*
Mrs. Blake, *Croydon*
Mr. Bidmead, *Academy, Hampstead*
William Boys, Esq. F. R. S. and F. S. A. *Sandwich*
Mrs. Boyd, *Margate*, 3 copies
Mrs. Boyd, *Sackville Street*
Edward Boys, Esq. *Salmston, Kent*
Mrs. Bouchery, *Swaffham, Norfolk*
Reverend William Boldero, *Woodford, Essex*
Mr. John Bonnycastle, *Royal Military Academy, Woolwich*
Bocking Book Club
Reverend Dr. Breedon, *Brese Court, Berks*
Mrs. Brock, *Much Euston, Essex*
Miss Braban, *Bryan House*
Henry Bridger, Esq. *Hythe*
Miss Brown, *Margate*
William Breese, Esq. *Crutched Friars*, 2 copies
Reverend Philip Brandon, *Deal*
Mr. Brooman, *Margate*
Mrs. W. Brooke, *Lincoln's Inn*
Mr. Brace, *London*
James Burne, Esq. *Custom House*
Mr. Burgess, *Ramsgate*
Mrs. Byrche, *Canterbury*

C.

Her Grace the Duchess Dowager of Chandos
The Right Honorable Earl Cowper, *St. John's College, Cambridge*
The Honorable Mr. Cooper, *St. John's College, Cambridge*
Mrs. Elizabeth Carter, *Deal*
John Carter, Esq. *Deal*

Mrs. Campbell, *London*

Robert Cawne, Esq. *Mercer's Hall*

William-Henry Cawne, Esq. *Christ Church College, Oxon*

Miss Frances Cawne, *Bryan House*

Miss E. Cawne, *Bryan House*

John Callanan, M. D. *Cork*

Reverend W. Chapman, *Margate*

Jonathan Chapman, Esq. *London*

Mrs. Chevely, *Queen's Palace*

Miss Chevely, *London*

Miss M. Chevely, *London*

——— Chuter, Esq. *Homerton, Essex*

Mrs. Chuter, *Homerton, Essex*

Mr. Clive, *Christ-Church College, Oxford*

Mr. Clutterbuck, *King's Street, Bloomsbury*

Ralph Clark, Esq. *Adelphi*

Mrs. Clark, *London*

William Colhoun, Esq. M. P. *Westminster*

Major Cook

Mrs. Cook, *Woodford, Essex,* 2 copies

Mrs. Cook, *Stratford, Essex*

Mr. John Cook, *Park Street, Grosvenor Square*

Miss Cook, *Bryan House*

Joseph Cotton, Esq. *Layton, Essex*

J. Cotton, Esq. *Exchequer*

Mr. Thomas Cotton, *Fenchurch Street*

Mr. Coxon, *Margate*

Edward Cotton, Esq. *Scotland*

John Cowell, Esq. *Margate*

William Crawford, Esq. *Cork*

Mrs. Crawford, *Cork*

Miss Crawford, *Bryan House*

John Crawford, Esq. *Crawford's Burn, Ireland*

Miss Crawford, *Ditto*

Arthur Crawford, Esq. *Craigavade, Ireland*

Mrs. Crespigny, *Camberwell*

Mr. John Cross, *Limehouse*

Cornet Carter, Oxford Light Dragoons

Mrs. Cartwright, *Grosvenor Street, Grosvenor Square*

D.

The Honorable Mr. Dillon, *Christ-Church College, Oxford*
—————— Dawson, Esq. *Amthill*
Mr. Davis, *Chancery Lane*
Reverend R. B. De Chair, *Shepherd's Well, Kent*
John Dickson, Esq. *London*
Miss Dickenson, *Queen Street, London*
Miss Dobbin, *Waterford*
Mrs. Downes, *Hereford*
Mrs. Draper, *Dublin*
Miss Dyer, *Homerton*

E.

Lady Eamer, *London*
Colonel Etherington, *Albemarle Street*
Captain Edwards, *Royal Navy*
Captain J. Edwards, *Ditto*
Captain Edwards, *Flintshire*
John Edwards, Esq. *Blackheath*
Mr. Edwards, *Trinity Hall, Cambridge*
George Ellis, Esq. *St. James's Place*
David-Mathias Erskine, Esq. *London*

F.

William Fermer, Esq. *Oxfordshire*
Peter Fector, Esq. *Dover*
Jesse Foot, Esq. *Dean Street, Soho*

G.

William Garrow, Esq. *London*
Thomas Garrett, Esq. *Nether Court*
James Gale, Esq. *Edgware Road*
Mr. Garner, *Margate*
George Gipps, Esq. *Harbledown, Kent*
Matt. Gilpin, Esq. *Temple*
Mr. Gilder, *Margate*
John Godley, Esq. *Dublin*
John Govett, Esq. *London*
Francis Gosling, Esq. *Lincoln's Inn*

John Goddard, Esq. *Woodford Hall*
Valentine Grimstead, Esq.
Silvanus Grove, Esq. *Woodford*
John Grove, Esq. *Ditto*
Miss Diana Grove, *Ditto*
Mrs. Grant, *Portman Square*, 3 copies
Mrs. Grosvenor, *Margate*
Mr. John Green, *London*
Mr. William Green, *Academy, Deptford*
Bartlett Gurney, Esq. *Norwich*

H.

The Right Reverend Dr. Horsley, Lord Bishop of Rochester.
Sir John Henniker, Bart.
Charles Hutton, L.L D. F.R.SS. Professor of Mathematics in
the Royal Military Academy, *Woolwich*
James Hamilton, Esq. *Margate*
Captain James Haverkam
——— Hadley, Esq. *London*
Miss Hartley, *Margate*
Mrs. I. Harvey, *Sandwich*
Mr. Haynes, *Cornhill*
Reverend Mr. Hayward, *Haverell, Suffolk*
Mrs. Hamilton, *Glocester Place*
Mrs. Hill, *Margate*
Mrs. Hill, *Hackney*
Mrs. Hill, *Stratford*
W. Hollingworth, Esq. *Admiralty*
Reverend Mr. Howley, *Winchester College*
Reverend John Hopkins, *Christ College, Cambridge*
John Houblon, Esq. *Hollingbury, Essex*
Captain Honyman, M. P. *Royal Navy*
T. Holmsted, Esq. *Bocking*
Thomas Hopkins, Esq. *Chigwell*
Thomas Hoggins, Esq. *Deptford*
Lieutenant H. Hughes, *Royal Navy*
Isaac Hutchinson, Esq. *Cannon Street*, 2 copies
Mr. Hunter, *Edinburgh*

I.

Captain Ievers

J.

Lady Jodrell, *Nottingham Place*
Miss Jackson, *Calcutta*
Miss H. Jackson, *Frances Street, Bedford Square*
Mr. Jarvis, *Margate*
Reverend R. Jeffreys, *Academy, Hall Place*
Edward Jenner, M. D. F. R. S. *Berkley*
Reverend A. Johnson, Vicar of *Baddow*
E. K. Jones, Esq. *Wanstead, Essex*
Mr. W. Jones, *Holborn*
Mr. T. Jones, *Ditto*
Mr. J. Jones, *Ditto*
Mr. Jones, *Mount Street, Berkeley Square*

K.

Honorable Lloyd Kenyon, *Christ-Church College, Oxford*
Frederick Kanmacker, Esq. *London*
Reverend Jarvis Kenrick, *Chilham*
Mrs. Kenrick, *Ditto*
William Kenrick, Esq. *Ditto*
Jarvis Kenrick, Esq. *Ditto*
Miss Kenrick, *Ditto*
William Kearney, Esq. *Waterford*
Mr. Kemp, *Trinity College, Cambridge*
Mrs. Kearsley, *Fleet Street*
Mr. Kearsley, *Ditto*
Mr. Kelly, *Academy, Finsbury Square*
Thomas Keith, Esq. *London*
Vice Admiral Kingsmill
Sámuel King, Esq. *Waterford*
———— Kirby, Esq. *London*
Mr. R. Kidd, *Margate*
Miss Kidman, *Margate*
Miss Knudson, *Hackney*
Miss E. Knudson, *Ditto*

L.

Reverend Mr. Lax, Professor of Astronomy, *Trinity College, Cambridge*

William Larkins, Esq. *Blackheath*

W. A. Latham, Esq. *Trinity College, Cambridge*

Mrs. Larking, *Clare House*

Reverend Mr. Layton, *Christ's Hospital*

George Leith, Esq. *Walmer, Kent*

Dr. Lettsom

Lieutenant Lewis

Mrs. Lee, *London*

Mrs. Le Geyt, *Canterbury*

Philip Le Geyt, Esq. *Magdalen College, Oxford*

Miss Le Geyt, *Bryan House*

Charles Le Bas, Esq. M. C. *Margate*

William Leighton, Esq.

———— Lean, Esq. *Hudson's Bay House*

Mrs. Leapidge, *Margate*

Captain Lloyd

M.

His Grace the Duke of Marlborough

Right Honorable Lord Montague, *St. John's College, Cambridge*

Honorable Mr. Marsham, *Christ-Church College, Oxford*

Reverend Dr. Maskelyne, Astronomer Royal

Francis Maseres, Esq. Cursitor Baron of the Exchequer

Lieutenant Marsh, *Royal Navy*

Mrs. Mathew, *Woodford*

Mrs. Mc Dougall, *Deal*

Mrs. Mackeson, *Deal*

Henry Mann, Esq. *Fenchurch Street*

Mrs. Mann, *Ditto*

Mr. Mayhew, *Ramsgate*

Hugh Mair, Esq. *London*

Miss Manson, *Portland Street*

John May, Esq. *Deal*

William Massie, Esq. *Southampton Street, Bloomsbury*

John Mc Connell, Esq. *Newland, Monmouthshire*

William Meredith, Esq. *Harley Place*
Mrs. Mills, *Mackington*
Mr. John Morris, *East Malling*
Richard-Hunt Micklefield, Esq. *Pamdon, Essex*
Captain Were Mudge, Royal Artillery, *Woolwich*

N.

Sir Simon Newport, *Waterford*
Lady Newport, *Ditto*
Reverend Mr. Nairn, *Kingston*
John Nash, Esq. *Salter's Hall*
George Neale, Esq. *London*
Mrs. Newport, *Waterford*
Abraham Newman, Esq. *London*
Thomas Nixon, Esq. *New Ormond Street*
Josiah Nottidge, Esq. *Bocking*
Mrs. Nottidge, *Ditto*
Thomas Nottidge, Esq. *Ditto*
Mrs. T. Nottidge, *Ditto*
John-Thomas Nottidge, Esq. *Trinity College, Cambridge,* 3 copies
Josias Nottidge, jun. Esq. *Bocking,* 2 copies
Reverend John Nottidge, Rector of *Aslington, Essex*
William Nottidge, Esq. *Russell Street*
George Nottidge, Esq. *Warmingford, Essex*
William Norris, Esq. *London*
William Nodes, Esq. *New Cavendish Street, Portland Place*

O.

Lady Oxenden, *Broome*
Miss Oakley, *Deal*
Reverend Charles Onley, *Stisted Hall, Essex*
Reverend John Ord, B. A. *Emanuel College, Cambridge*
Miss Otte, *Homerton*
Miss Owen, *Norfolk Street*

P.

Lady Prescott, *Portman Square*
William Parnell, Esq. *Trinity College, Cambridge*
I. G. Palairet, Esq. *Epsom, Surry*
Reverend George Paroissier, *Hackney*

Mrs. Palairet, *Berner's Street*
Mrs. Parkinson, *Tooting, Surry*
Miss Petrie, *Chelsea*
———— Peake, Esq. *London*
Mr. Personneux, *Margate*
Leonard Phillips, jun. Esq. *Scotland Yard*
———— Postletwhait, Esq. *London*
Mrs. Poole, *Woodford*
John Poole, Esq. *Teddington, Middlesex*
Mrs. Prestwidge, *Mincing Lane*
Miss Prestwidge, *Ditto*
Captain Price, *Betsanger*
Mrs. Purdew, *Deal*

R.

The Right Honorable Lord Rivers
The Honorable Colonel Rawdon
The Honorable Henry Rider, *St. John's College, Cambridge*
Reverend Matthew Raine, B. D. Head Master of the Charter-
    House School
Jonathan Raine, *Charter House*
Miss Ralph, *Woodford*
———— Reynolds, Esq. *London*
William Robinson, Esq. *Woodford*
Edward Robarts, Esq. *Ealing, Middlesex*
Mrs. Rook, *Stratford, Essex*
Christopher Rolleston, jun. Esq. *Denmark Hill*
Joseph Ruse, Esq. *Ramsgate*
Isaac Rutton, Esq. *Uspringe Place*
Miss Rutton, *Ditto*
I. C. Ruding, Esq. *Francis Street, Bedford Square*

S.

The Right Honorable Lady Sondes, 2 copies
Sir George Shuckburgh-Evelyn, Bart.
Mrs. Sambler, jun. *London*
William Sambler, jun. Esq. *Ditto*
Jacob Sawkins, Esq. *Margate*
Mrs. Sawkins, *Ditto*

Reverend Charles Sawkins, *Christ-Church*, *Oxford*
Miss Sawkins, *Margate*
Mrs. Savill, *Stisted Hall*, *Essex*
Mr. Saffery, *Margate*, 2 copies
Mr. George Saffery, *Ditto*
Mr. Sanneville, *Margate*
Miss Scott, *London*, 3 copies
Mrs. Scratton, *Sutton Hall*, *Essex*
J. Scratton, Esq. *Hackney*
W. Seward, Esq. *Great Portland Street*
Baker-John Sellon, Esq. *London*
———— Sermon, Esq. *London*
S. Shelly, Esq. *George Street*, *Hanover Square*
Miss Short, *Great Ormond Street*, 2 copies
Mrs. Shum, *Bedford Square*
Captain Sharp, *Dover*
———— Shirley, Esq. *Stratford*, *Essex*
Mr. Shirriff, *Deptford*
Mr. Samuel Silver, *Margate*
Mr. Slater, *Margate*
Lieutenant Smith
Miss Smith, *Stratford*, *Essex*
Miss Clarinda Smith, *London*
Mr. Sotheby, *York Street*
Mr. S. Sotheby, *Ditto*
Miss M. A. Sotheby, *Bryan House*
Miss Marian Sotheby, *York Street*
George Stratton, Esq. *Devonshire Place*
Mrs. Stringer, *Peckham*
———— Stringer, jun. Esq. *Ditto*
Mr. William Starkey, *London*
Lieutenant Stockford
Mrs. Nicholas Styleman, *Norwich*
Richard-Joseph Sulivan, Esq. M. P. *Grafton Street*

T.

Mr. Tavel, M. A. Fellow of *Trinity College*, *Cambridge*
Mrs. Tambs, *Sandwich*, *Kent*
Mr. Tallbot, *Stratford*, *Essex*

Miss Thrale, *Great Cumberland Street*
Mr. Thompson, *University College, Oxford*
Robert-Sterne Tishe, Esq. *New Park, Wincanton*
Robert Tindal, Esq. *Chelmsford, Essex*
C. Tower, Esq. *St. John's College, Cambridge*
John Toke, Esq. *Canterbury*
Robert Tournay, Esq. *Hythe*
Mrs. Tournay, *Ditto*
Mrs. Tresilian, *Margate*
Miss Troward, *Margate*
Reverend Dr. Turner, Dean of *Norwich*, and Master of *Pembroke Hall, Cambridge*
Alexander Tulloh, Esq. *London*, 2 copies
Miss Tulloh, *Ditto*
John Tuck, Esq. *London*
Mrs. Twhaytes, *Fenchurch Street*
Mrs. John Twining, *Twickenham*
Thomas Tweed, Esq. *Sandon, Essex*
Mrs. Tyson, *London*

V.

Reverend Dr. Valpey, *Reading*
John Vieusseux, Esq. *Baddow, Essex*
Major Vaughan

W.

The Right Honorable William Windham, Secretary at War
Reverend Dr. Wakeham, Dean of *Bocking*
Reverend Henry Wakehan, *Bocking*
Luke Walford, Esq. *London*
Mr. W. Walker, *London*
Wakeland Welch, Esq.
Mrs. Welch, *Maryland Point*
Mr. Were, *Margate*
———— Whitby, Esq. B. A. *Trinity College, Cambridge*
———— Whatman Esq. *Trinity College, Cambridge*
Mrs. Whitehead, *Margate*
Lieutenant Colonel Wilson
Miss White, *Jeffries Square*
Reverend Samuel Wix, A. B. *Christ College, Cambridge*

William Wilcox, Esq. *Hodsdon, Herts*
Edmund Wilcox, Esq. *Ditto*
Mrs. A. Wilcox, *Ditto*
Miss Wilkins, *London*
Mrs. Witherden, *Ramsgate*
Colonel Wrench
Mrs. Wyatt, *Bocking*

Y.

John Yenn, Esq. R. A. *Kensington*
Miss York, *Ely Palace*

Z.

Miss Zurhurst, *London.*

[xxv]

CONTENTS

OF THE

SCIENTIFIC and HISTORICAL PARTS of the LECTURES.

LECTURE I.

Page

On Optics — — — — — — — — — 4

On the progressive Motion of Light, and its Velocity — — 5

The Nature of Light.—Of Refraction and Reflection — — — 7

The Laws of Refraction and Reflection — — — — 10

Of the Nature of Lenses — — — — — 14

Of Gravity — — — — — — — — 16

Of Pendulums — — — — — — — 18

Of the Increase of Matter about the Equator, and that it is occasioned by the Centrifugal Force arising from the Motion of our Earth round its Axis — — — — — — — — 20

LECTURE II.

Origin of Astronomy — — — — — — 23

First Measure of Times and Seasons — — — — — 26

Symbolical Writing the Origin of Idolatry — — — — 30

History of the Improvements in Astronomy.—Of the Advantages resulting from the Investigation of this Subject — — — — 31

The Appropriation of the Letters of the Greek Alphabet, to express the different Sizes of the fixed Stars — — — — — 41

Of the Reformation of the Calendar by Julius Cæsar — — 42

A

Page

The Old Calendar abrogated, and a New One established by Pope Gregory XII.   43

Of the Dominical or Sunday Letter   —   —   —   —   —   44

Of the Introduction of Telescopes   —   —   —   —   —   47

L E C T U R E   III.

Introductory Address   —   —   —   —   —   —   49

Of the Figure and Affections of the Globe we inhabit; and of the Appearances of the Heavens and Heavenly Bodies in consequence of the latter   50

Of the Atmosphere   —   —   —   —   —   —   54

Of the Magnet   —   —   —   —   —   —   55

Of the Horizon   —   —   —   —   —   —   57

Of the Meridian of the Zenith and the Nadir   —   —   —   —   58

Of the Equator.—Of the Sensible and the Rational Horizon   —   —   59

The Celestial Globe and Armilary Sphere explained, with their Uses in referring to Celestial and Terrestrial Phœnomena   —   —   60

Of the longest and of the shortest Days in both Northern and Southern
Latitudes   —   —   —   —   —   —   —   —   63

Of the Latitude and Longitude of the fixed Stars; also of the Planets.—The
Difference between the Longitude and Latitude of Places on our Earth,
and those of the Stars   —   —   —   —   —   —   64

Of Declination.—The Tropics, why so called.—Of the Sun's apparent Motion North and South   —   —   —   —   —   65

Of Zones; and the different Lengths of the Days and Nights at each of them   66

Of the Climates; and the Length of the Natural Day at each of the Twenty-
Four Circles parallel to the Equator   —   —   —   —   68

Of a right, a parallel, and an oblique Sphere   —   —   —   69

Of the Arctic and Antarctic Circles   —   —   —   —   —   70

Of the Precession of the Equinoxes   —   —   —   —   71

Of the Inequality of the Sun's apparent Motion.—The Effects of the Sun's
Rays falling with greater or less Obliquity on our Earth   —   —   74

L E C T U R E   IV.

The Copernican System explained.—Of the supposed Nature of the Sun   80

Of Mercury   —   —   —   —   —   —   82

Page

Of Venus — — — — — — — 83

Of the Earth. — Of Mars — — — — — — 84

Of Jupiter — — — — — — — 86

Of Saturn — — — — — — — 88

Of the Georgium Sidus — — — — — — 91

Of the Comets — — — — — — — 92

Of the Quadrant and its Uses — — — — — 97

Of the Adjustment of the Quadrant — — — — 99

LECTURE V.

Of the Manner in which Mathematicians and Astronomers estimate the real
Sizes and Distances of the Heavenly Bodies belonging to the Solar
System. — Of the Diminution of Heat and Light. — Of the apparent
Diameters of Bodies according to their Distances — — 103

Of the true and apparent Places of the Heavenly Bodies — — 106

Of Aurora and Twilight. —The Effects of the different States of the Atmosphere — — — — — — — — 107

The Sun and Moon both seen before they rise above the Horizon — 108

Of Parallax — — — — — — — — 109

Of the different Parallaxes — — — — — — 110

Parallax not referable to the fixed Stars — — — — 111

Of the Transit of Venus — — — — — — 113

Of the apparent Size of the Sun and of the Moon — — — 116

Of the Aberration of Light — — — — — — 117

LECTURE VI.

Of the fixed Stars; and a View of that sublime Field of Investigation, the
Universe.—Why we suppose the Stars to be fixed in a concave Sphere 125

Of the Aurora Borealis — — — — — — 130

Changes in the fixed Stars — — — — — — 132

Of Motion and Gravity — — — — — — 134

LECTURE VII.

Page

The Terms Gravity, Weight, Attraction, and Centripetal Force, how applied to express the same Thing — — — — — — 143

Of the Orbits of the Planets — — — — — — 144

Of the Nature of an Ellipse — — — — — — 145

Why the Planets revolve in elliptical Orbits — — — — 146

How the Motions of the Planets are perpetuated.—Of accelerated and retarded Motion — — — — — — — 147

Of the Centrifugal and Centripetal Forces.—Of the Motions of the Satellites 148

Of the longer Axis of an Ellipse, or the Line of the Apses — — 149

Why the Motion of a Planet is swiftest when it is Perihelion, and slowest when at Aphelion — — — — — — — 150

Of the mean Distances of the Planets from the Sun.—The Annual Motions of the Planets how ascertained — — — — — 151

Of the apparent diurnal Motion of the Heavenly Bodies.—Of their apparent Retrograde Motion — — — — — — 152

A Description of the Eye, and its Capacities; also its Effects explained 153

LECTURE VIII.

A Description of the Orrery in its divisible State, and of the Motions of the Planets of our System, elucidated by the Application of the Planetarium Part of it; also by Diagrams, and the known Laws of Projectile and Centripetal Forces — — — — — — — 165

The diurnal Motion of the Planets how ascertained.—Why all loose, globular Bodies turning on their Axes, must be oblate Spheroids.—Why all the Heavenly Bodies appear equi-distant from us — — — 169

The Order in which the Stars appear after Sun-set — — — 170

How Sir Isaac Newton estimated the Heat and Light received by each Planet of the Solar System — — — — — — 171

The relative Bulks of the Sun and Planets.—Why the Sun appears to pass through all the Signs of the Zodiac in One Year — — — 173

CONTENTS.

Of the true or direct Motion of the Planets; or, in consequentia. — Of their
apparent or retrograde Motion; or, in antecedentia. — The Times of the
Retrogression of a superior and of an inferior Planet different. — The
Times of each Planet being retrograde — — — — 175

When a Planet is said to be in Conjunction with the Sun. — Of true and
apparent Conjunction. — Of the relative Situations to each other in
different Parts of their Orbits — — — — — 176

Of the Aspects of the Planets, and how expressed. — Of the inferior and of
the superior Conjunction of Venus and Mercury; and of the greatest
Elongations — — — — — — 177

Of the direct and retrograde Motion of Venus — — — — 179

When the Motion of an inferior Planet is direct, and when retrograde. — The
Absurdity of the Ptolemean System evinced — — — 181

Of the Phases of the Planets — — — — — — 182

The Time of a superior Planet being retrograde — — — — 183

The Truth of the Copernican System demonstrated — — — 184

LECTURE IX.

Of the Obliquity of the Axis of the Earth to the Plane of its Orbit — 187

Mr. Norwood's Mode of ascertaining the Number of Miles contained in a
Degree of the Earth; and the Improvements made since his Time in
this Calculation, by a Knowledge of the true Form of the Earth on the
Principles of Gravity — — — — — — 189

The Nutation of the Earth's Axis accounted for — — — 191

Of Solar and Sydereal Time, and of the Inequalities of the Solar Days
amongst themselves — — — — — — — 192

Of the Tropical or Civil Year. — — — — — 198

Of the Vicissitude of the Seasons, and of the different Lengths of Day and
Night throughout the Year — — — — — — 199

Of the Motions of the Moon; her different Phases; and the Inclination of
her Orbit to the Ecliptic — — — — — — 203

Of the Variety in the Moon's Affections — — — — 204

Of the Telescopic Appearance of the Moon — — — — 207

Of the Lunar Synodical, and the Lunar Periodical Month — — 209

L E C T U R E X.

Page

Of the Laws of Shadow — — — — — — — 217

Of Eclipses of the Moon — — — — — — 218

Of Solar Eclipses — — — — — — — 220

Of the Moon's Atmosphere — — — — — 221

Of the Periods of Eclipses — — — — — — 222

The Phœnomena of the Tides considered; and how produced — — 224

Of Winds; particularly that called the Trade Wind — — 231

Elements of Trigonometry — — — — — 237

Trigonometrical Problems — — — — — 244

The general Principles of a Celestial Globe explained — — 253

Celestial Problems — — — — — — 255

Explanation of the Tables used in White's Ephemeris — — 271

Characters of Bodies belonging to the Solar System, and of the Constellations of the Zodiac.—Characters of the Affections of the Planets, and their Significations; also other Characters explained that are used in the Ephemeris — — — — — — — 277

A View of the Solar System; referred to Plate 6 — — 278

Catalogue of the Constellations in the Northern Hemisphere; referred to Plate 4 — — — — — — — 280

Names of the Stars of the firſt Class in the Constellations; referred to Plate 4   282

Catalogue of the Constellations in the Southern Hemisphere; referred to Plate 5 — — — — — — — 283

Names of the Spots in the Moon; referred to Plate 13 — — — 285

A Vocabulary of Terms — — — — — 287

E R R A T A.

Page.	Line.	
6	18	After light, dash improper
14	16	(for) these glasses (read) the glasses
17	2	(for) the absence of which (read) the influence of which
21	2	(for) those effects (read) these effects
36	24	(for) revolutions (read) of revolutions
43	12	(for) every every (read) every
65	28	(for) between north and south (read) north and south
68	5	(for) they divide it into 24 parts, 12 on each side of the equator (read) they divide it into 24 parts on each side of the equator
70	19	(for) longer their nights (read) longer than their nights
86	7	(for) of the system (read) of it
89	10	(for) place (read) plane
101	22	(for) set no value (read) therefore setting no value
122	4	Poetry, (for) glass (read) gloss
161	8, 19, 21, 24	(for) focusses (read) foci
166	9	(for) terminate (read) terminator
166	23	(for) there (read) then
173	8	after times (read) as large as our Earth
173	13	(for) explain each (read) explain them
173	15	(for) each circumstance (read) the circumstances
177	6	(for) 180 (read) 120
178	8	(for) distance (read) distant
180	3	now is superfluous
186	15	(for) their (read) there
188	4	(for) those which are (read) those being
188	5	(for) turned on its orbit (read) turned on in its orbit
188	23	(for) orth at (read) or that
190	4	(for) precession (read) precision
197	12	(for) eclipse (read) ellipse
206	15	(for) arallel (read) parallel
209	28	(for) of performing (read) of her performing
219	29	(for) orbits (read) orbit
233	18	(for) virtues, let (read) virtues, which let
233	19	(for) They (read) will
233	20	(for) will render (read) render
240	16	(for) or semicircle (read) or a semicircle
240	20	(for) on its orbit (read) or its orbit
240	25	(for) distances (read) distance
241	22	(for) a local use of the quadrant only (read) a local use only of the quadrant
241	27	(for) size and distance (read) sizes and distances
242	28	after equator (read) with the horizon of places in about 52° of northern latitude
244	25	(for) as are (read) are
246	5	(for) the horizon (read) it
248	12	(for) two parts (read) two places
253	6	omit and
262	1	(for) on your meridian (read) or on your meridian
262	20	(for) globes (read) globe
263	17	(for) westward (read) western
297	28	(for) they are not (read) they are not so

Plate I.

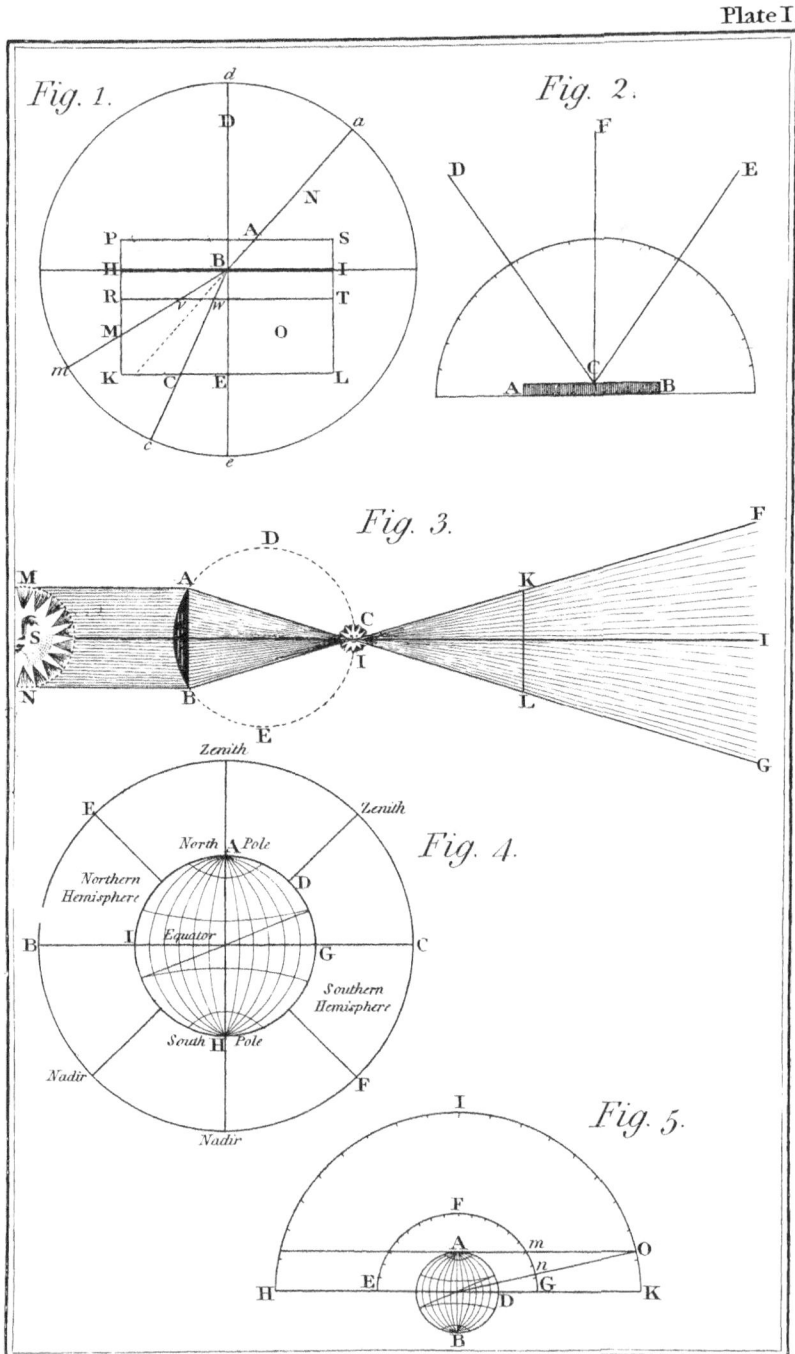

Fig. 1.

Fig. 2.

Fig. 3.

Fig. 4.

Fig. 5.

M. Bryan delin.ᵗ

T. Conder sculp.ᵗ

A

COMPENDIOUS SYSTEM

OF

ASTRONOMY.

LECTURE I.

GENERAL INTRODUCTORY REFLECTIONS ON VARIOUS SUBJECTS.
ON OPTICS.—THE EYE.—LIGHT.—REFRACTION.—RE-
FLECTION.—GRAVITATION.—PENDULUMS, &c.

NATURAL objects, when properly contemplated, continually admonish us in the important science of Divine Wisdom, leading us to consider our situation in this sublunary state, our connections and dependencies;—from which we learn the duties required of us, and the exertions we are capable of making.

From the consideration of our mental faculties, we infer the exalted idea of a future state of existence, so naturally rising in the intelligent mind; which reflects on the never-ceasing energy of the mental power, and its independency of all mortal circumstances.

Thus perceiving what is the purer essence of our nature, and what the grosser, we are conscious that our present existence was not the primary or principal intention of our Creator: yet, as it is allotted us preparatory to that for which we were created, it claims our particular attention; becoming either advantageous to us or otherwise, accordingly as we deal with the objects which surround us.

B

Our superiority in the scale of beings, gives us the power of applying to our own use, the gifts of Providence by which we are surrounded, with the greatest advantage, not only so as to supply the necessities of our mortal nature, but also to derive considerable mental gratification from them.

Shall we then neglect rightly to use the gift of reason, and thereby become unworthy of such a boon, as well as lose all the benefits to be derived from it? Certainly not. Let us rather, on the contrary, so exercise and improve our understanding, as to form a right judgment of the value of things, by which alone we can be enabled to conduct ourselves according to the proper circumstances of the state in which we are placed: a business which requires more caution in the investigation than young minds are apt to imagine;—implying a thorough knowledge of the human mind, which can be obtained only by a careful examination of its capacities and infirmities.

Before entering upon this important investigation, my duty prompts me to offer some reflections on that Hand which formed us,—that Divine Mind which directs all our involuntary operations,—and that Benevolence which renders these operations instrumental to the comfort and happiness of all its creatures.

Yet, how can I presume to recount the works of the Almighty, or shew the wisdom of his counsels! Far above the narrow scale of human enquiry, far out of the reach of the feeble efforts of human comprehension, are such investigations: yet are his attributes discoverable in his wise administration, made evident to us through the medium of our senses.—Let us then receive these emanations of the Divine Mind shed down upon us, with as much joy and thankfulness as we do the rays of the sun when falling upon our crops of corn, making them yield forth their abundance.

This world is by no means barren of comforts to those who cultivate a relish for the delights it offers, avoiding satiety; for by a proper application of the objects of sense, we shall learn how to render the things of this life not only serviceable but delightful to us.

If we attentively regard the wonders of creation, we shall discover gradations; some rising superior to others in that excellence, of some kind or other, which is peculiar to each particular species. In vegetable and mere animal nature, this is unavoidable; but in the higher class of animated nature, in man, much depends on his own exertions, as is evident from the instances of the learned men of all ages and all climates; some of whom, though born under all the disadvantages of superstition and barbarifm, yet by a right application of their understanding, they have signalized themselves in the delightful fields of science and virtue, not excelled even by those who have lived and studied in the most civilized nations with all the advantage of instruction and method to regulate their researches:—and this excellence the former have attained to by a due cultivation of their minds, by which they have increased in knowledge, by regular gradations, till they have arrived at the highest pitch of mental improvement.

The operations of the mind distinguished by logicians, are four.

The first is perception, from which refult our ideas.

The second, judgment; which is the assembling ideas together, and comparing them.

The third is reasoning, which is the exercising our minds in producing proofs to apply to the discovery of facts.

The fourth, which is the last operation of the mind, is called method;—as we must perceive, judge, and reason, before we can methodize. The mind must be stored with the knowledge resulting from the foregoing operations, before it can be capable of dispofing its intellectual acquirements into classes, or of uniting them according to their proper connections and dependencies.

How delightful is the task of enquiry! how important the advantages resulting from investigation! Amply is the searcher into nature rewarded, by the extention of ideas and the strength of judgment acquired; by which the human understanding is enabled to soar above vulgar prejudices, and to view the works of God with satisfaction,—deriving consolation from every object in nature.

Suppose we employ the operations of the mind at present on an object which aptly offers itself; being so connected with the subjects of these Lectures, that, without it, we could never have contemplated the wonders of those regions they treat of—namely, the organ of sight. Do we not perceive its fitness to the office assigned it? and do we not from thence infer, the power and wisdom of that Being who constructed it? " He that made the eye, doth not he see?" Thus proceeding, according to the natural operations of the mind, we discover not only its use, but the necessity of such an organ to rational beings, as well as to almost all animated nature, capable of self-action; as, without it, how could existence have been continued? how could the latter have sought their food, or the former have prepared it? From this evidence of God's preventing goodness we naturally infer, that every part of his creation is replete with the like instances of benevolence in their several dispensations.

In order to your understanding the wonderful contrivance manifested in this little organ, to fit it for the rays of light, it will be necessary to contemplate the nature and properties of that subtile fluid.

The particles of light are inconceivably small, and their velocity exceeds human conception. The laws by which they are governed, and by which other bodies act upon them, by reflections and refractions, without altering their original properties, and the facility with which they insinuate themselves into substances of the compactest texture,—are circumstances which excite astonishment in the most cultivated minds.

The Sun is the fountain of heat and light, which are kindred properties, only differently modified; sometimes they act unitedly, and sometimes independantly of each other.

This fluid is universally diffused, and perpetually flowing from its source, the sun. Mixing with our atmosphere, it supports animal and vegetable existence, adorns all nature, and is the first natural principle of motion and vitality.

That light is a material substance, has been proved by its being subject to those laws which characterize materiality; such as decomposition, force, and being prevented from pursuing its strait course by the interposition of an opake body. Its natural motion, like that of all bodies, is uniformly progressive, as has been proved by many experiments; also that its velocity is the same, whether original, as from the sun, or reflected, as by the planets.

From observations of the eclipses of the satellites of Jupiter, it has been discovered, that light is propagated from the Sun to our Earth in 8 minutes 12 seconds. These eclipses, when they happen at the time the Earth is interposed between the Sun and Jupiter, are observed about 8 minutes 12 seconds sooner than they should do by the tables; and when the Sun is between the Earth and Jupiter, they happen 8 minutes 12 seconds later than they are calculated for. These circumstances are owing to the light of the satellites having further to come in one case, than in the other, by the whole diameter of the Earth's orbit.

These observations clearly prove the progressive motion of light, and ascertain its velocity, as these inequalities are certainly occasioned by the time which the light takes in traversing that quantity of space, which is equal to the diameter of the Earth's orbit, and not to any inequality in the motions of the satellites in their orbits; for the same happens in all parts of their revolutions.

Although it is impossible to convey an adequate idea of the amazing velocity of light, yet, that we may form some conception of it, I shall just observe, that it moves at the rate of twelve millions of miles in a minute; as is known, by comparing the Earth's mean distance from the Sun with the foregoing phœnomenon.

The unlearned in the sciences are astonished at the minute calculations made by mathematicians, of the distances, sizes, &c. of the heavenly bodies, and of our Earth; because, they are not acquainted with the principles of this science. I hope in its proper place to convince my pupils, that such calculations may be accurately made by those versed in the mathematical sciences.

I shall not enlarge on the subject of optics, or the nature and properties of light, in these lectures, more than will be necessary to your conceiving the nature and extent of celestial observations, after having recalled the attention of my dear pupils to a retrospect of those clear evidences they have perceived of this admirable subject, in my lectures on optics—The wonderful organization of the eye! the uses of its different parts, and the delight it affords.

The rapture with which we have contemplated the separated rays of light,—in all their distinct and glowing colours,—when charmed with the variegated hues, we have almost forgotten what they were to authenticate,—till anxious for conviction we have returned to the theory,—when, on comparing the admirable experiments of Sir Isaac Newton, with his inferences, we have been doubly gratified by truths so clearly elucidated.

Then, on comparing the properties of light with the mechanism of the eye, we have perceived that the eye was formed for the rays of light—that it had the power, in a great degree, of adjusting itself to their strongest and weakest impressions—that its humors were adapted by their powers of uniting, refracting, and reuniting, to answer, in the most perfect manner, the purposes of vision.

Then considering the powers employed externally to aid this delicate machinery, such as the muscles, by which all its motions are regulated, and that each of these has its particular office assigned it;—one raising and supporting the eye, another lowering it;—one drawing it towards the nose, another towards the temple, another making it roll about with ease, and the last moderating the movements of all the rest, so that they should not exceed certain limits—how has the prescient and preventing providence of God, thus manifested to us, filled our hearts with love and admiration!—how sensibly have we felt this conviction of an all-wise, all-seeing, and all-bountiful Creator!

In proceeding with the subject which occasioned this digression, I shall endeavour to be as concise and intelligible as possible, avoiding all superfluous illustration and abstruse terms of science.

By a ray of light, is meant the least particle of light that can be separately impelled.

By a pencil of rays, a parcel of rays issuing from a point, or proceeding towards one.

When an object is illuminated by the sun, or a candle, every physical point on the surface of that object which is turned towards the light, sends forth rays in all directions, by which power it is, that it becomes visible to us.

That side of the object which is turned from the light, may also be seen, but it appears dimmer, that part not receiving the direct influence of the light, but receiving it by reflection only from surrounding objects. All objects would be invisible on the side turned from the radiant point, if it were not for that property of light which causeth it to be reflected.

Before the admirable Newton, by his judicious and well conducted experiments, discovered that the particles of light were turned back, before they touched the reflecting surface, it was generally supposed that the rays of light were reflected by striking against the impervious parts of solids, as other bodies are; but that great and penetrating genius discovered the error of this supposition, and has taught us, that the matter of light coming in contact with some unknown subtile medium, which is equally diffused over the whole surfaces of those substances, and striking it with a determined force, is thrown back by the re-action of this fluid medium; which fluid he conceives to be subtler than air, or light, and an intermediate agent.

" If light, says this great man, were reflected by striking on the solid parts of bodies, their reflections would not be so regular as they are found to be: as, however polished the smoothest surface may appear to our sight, or feel to our touch, yet is it an assemblage of inequalities; for in polishing glass with sand, it is not otherwise polished than by bringing its surface to a very fine grain, and from such a surface, the small particles of light would be as much scattered as from the roughest."

This conclusion is so convincing, that although we cannot discover the invisible agent which causes reflection, yet we must admit the probability of its existence.

That there is such an active medium, whose agency cannot be perceived but by its effects, we are certain, as, without allowing that there is, how can we account for all the effects we call gravity? of which I shall say more hereafter.

That the rays of the sun, or a candle, proceed in strait lines, whilst moving in the same medium, is evident by the shadows cast by opake bodies; but that in falling obliquely from one medium on to another, of a different density, they are refracted, is proved by experiments.

By a medium, is meant any transparent body which suffers light to pass through it, as air, glass, water, &c. These mediums differ in density, therefore the light, in passing from one to the other, is refracted; and the deviation is in some proportion to the difference of density of the two mediums.

The rays of the sun, in passing out of the ethereal region into our gross atmosphere, are refracted, so that we see that luminary before it has risen above our horizon, of a morning, and after it has sunk below it, of an evening; so that by this refractive power of our atmosphere, we enjoy the light of the sun a considerable time longer than we otherwise should do, and escape the disagreeableness of a sudden transition from one extreme to another.

The rays of light which fall perpendicularly on a different medium are not refracted, as when they fall obliquely; and when the sun is in the zenith of any place, that spot to which it is vertical, receives its direct influence, and suffers the extremest heat, receiving a greater quantity of the sun's rays, in proportion to the space.

The refraction which light is subject to, is easily seen in a familiar instance, by placing any object at the bottom of a bason, and withdrawing yourself from it, till it is lost to your view, when, if another person pours water into the bason, it will become visible to you, even in the very situation which before rendered it invisible. This is the consequence of the rays of light being refracted, which causes the image of the object to appear higher in the bason.

I think it proper to acquaint you with the nature and limitations of the laws of refraction and reflection, although it is not necessary to these Lectures to go into a close investigation of those laws.

Sir Isaac Newton's Theory, as I comprehend it, supposes all bodies to be filled with a subtile fluid, and that its action extends all over, and to some distance from their surfaces; and that it is the action of this fluid which causeth refraction, and the re-action of it, when it receives a given impulse, that causeth reflection. Thus the same fluid, by its different mode of action, is supposed to transmit and reflect light, and the latter, apparently from the substances themselves.

This subtile fluid, that great man conceives also to be the cause of gravity or attraction, and that it exists in all space; that this medium is condensed in our atmosphere, and still more so in the substances on our earth; and that its density is in proportion to the number of particles in the substances it penetrates. That if we admit such a fluid to be diffused through imaginary space, it must be in its natural state more rare than we can conceive, otherwise it would resist the motions of the planets very considerably; whereas the contrary being the case, as is evident by the regularity of their motions, we may be certain that this fluid must be very rare, so rare indeed, that it freely penetrates the pores of glass and the densest bodies.

Refraction and reflection are distinct properties, yet are they analogous, as being the operations of the same agent, this subtile medium.

The general laws of refraction are, that a ray of light, in passing out of a rarer into a denser medium, is refracted towards the perpendicular, that is, towards the axis of refraction, and hence the angle of refraction is less than the angle of incidence; whereas those two angles would have been equal, had the ray not been refracted. The physical cause of this may be, that the attraction of the denser medium, acting perpendicularly to the oblique direction of the incident ray, diverts the ray out of its course, in proportion to the superior force it exerts. But when a perpendicular ray falls upon a denser medium, in that case it suffers no refraction, because (we may still suppose) the attracting power, acting in the same direction with the ray, cannot cause it to deviate from its strait course.

On the other hand, when a ray of light passes obliquely from a denser to a rarer medium, it is refracted from the perpendicular; so that the angle of incidence is less than the angle of refraction.

I shall endeavour to explain the laws of refraction upon Sir Isaac Newton's principles of attraction, referring to *Dia. fig.* 1. *plate* 1. Suppose the line H I to be the boundary of two mediums, air and water, N and O; the upper one N the rarer, as the air; the lower one O the denser, as water; which latter is included between the lines RT, KL. The attractions of the mediums will be in proportion to their densities; that is, the compacter or heavier medium possesses a greater quantity of attracting particles, or rather, suffers a greater accumulation on its surface, of that subtile medium which causes gravity or attraction.

Suppose PS to be the distance to which the attracting force of the denser medium extends itself within the rarer. Suppose the line from A to represent a ray of light falling obliquely on the boundary of the mediums, which is represented by the line HI, or rather upon the line PS, as it is there that the denser medium begins to act. As all attractions are performed in lines perpendicular to the attracting body, when the ray arrives at A, it will begin to be turned out of its strait course by the superior force by which it is attracted by the medium O, more than by the medium N, which force draws it towards the perpendicular of the surface O. The ray is bent from its strait course continually all the way in every point of its passage between PS and RT, within which distance the attraction acts. Between these lines it describes an arch, but beyond RT, being out of the sphere of the attraction of the medium N, it proceeds uniformly in a right line from the point W to C.

Now let us reverse the experiment, by supposing N the denser and more attracting medium, and O the rarer, also HI the boundary of those two mediums, as before, and RT the distance to which the upper, and now the denser medium, exerts its attraction within the rarer; even when the ray has passed the point B, it will be within the sphere of the superior attraction of the upper and denser medium. Now, as the attraction acts in lines perpendicular to the surface of the upper medium, the ray will be continually drawn upwards, toward the denser medium, and from the perpendicular of the lower or rarer medium; that is, from its strait course BK, towards H I, in the direction B V. The two forces acting upon it, cause it to have a compound motion, whereby it describes a curve from B to V. When it has arrived at V, being out of the influence of the medium N, it will again go on in a right line, from the end of the curve to M.

The attraction of the medium N diminishes continually, as the ray proceeds from B towards the limit of superior attraction R T, because fewer and fewer parts continue to act with superiority. In this diagram, when the ray of light is supposed to fall out of a rarer on a denser medium, being refracted towards the perpendicular of the denser medium, the angle of refraction C B E is less than the angle of incidence A B D. When we suppose N to be the denser medium, and O the rarer, and the ray to be passing out of the denser medium N into the rarer medium O, the refraction being still towards the denser medium, the ray is turned from the perpendicular of the rarer medium, and therefore the angle of refraction E B M, is greater than the angle of incidence A B D; as may be seen, by transferring the angles to the correspondent letters on the circle.

All the rays of light which proceed from a large body at a very great distance, may be considered as parallel rays, as the greater the distance, the nearer they approach to a parallel direction.

As a luminous body may be seen from all points to which a strait line can be drawn, a luminous object must, by some unknown power, return rays in all directions; the effect of this power is called reflection.

The reflexibility of light, or the manner in which it is thrown back into the same medium through which it is transmitted, is subject to the same mechanical law as regulates the reflections of all bodies, the angle of reflection being always equal to the angle of incidence. This is easily seen by letting a ray of light fall upon a mirror; when we shall perceive, that the inclination of the reflected beam will be exactly equal to that of the incident one; and consequently the angles made by each will be the same.

Suppose A B, *fig.* 2, *plate* 1, to be a mirror, which receives a ray of light D C. This being the incident ray, and C E the reflected one. By comparing the degrees cut on the semicircle by the two rays, you will perceive them to be the same; and therefore the angle which the reflected ray makes with the perpendicular C F, is equal to the angle made with C F by the incident ray, that is, the angle D C F equal to the angle E C F.

Having informed you of the general laws of reflection and refraction, and explained, as well as I am able, the cause of those effects we perceive of them, I shall next proceed to inform you of the degrees of the power of light, under different circumstances and modifications.

Light radiating from a centre, diminishes as it diverges; because, by filling a larger space, it becomes more and more diluted. Thus it is easy to calculate the intensity of heat or light from a centre, if we know what the degree of heat or light is at that central spot, or comparatively to judge of its diminution according to its distance from the centre; light and heat diminishing as the squares of the distances increase. Thus, at two rods distance, the heat or light is four times more dilute or less intense than at one rod.

The particles of light are not all of the same size, and therefore are not all refracted in the same degree by the same power applied, the largest particles, by the momentum of their motion, being the least liable to be turned from their ftrait course, and the smallest the most easily; as all bodies require to be acted upon by forces proportioned to their magnitude, in order to give them a determined direction: the different sizes of the particles of light occasion, by their impressions, those diversities of tints exhibited by the prism. This instrument, by its peculiar construction, is calculated to divide or separate the rays of light from each other, according to their gradations of colour and size: the smallest or most refracted rays, exhibit violet colour; the next in size, being less refrangible, exhibit indigo; the next blue; the next yellow; the next orange; and the next red, which last consists of the largest particles of light, and of course the least refrangible.

If in the rays of light, separated by the prism, we place a double convex lens, they reunite at its focus, forming a spot of perfectly white light. If we ſtop one of the separated rays, and prevent its uniting with the others, the spot at the focus of the lens will not be quite white, but made up of a mixture of all the other colouring particles, exhibiting that hue which such a combination is known to produce. Thus clear and easy are all the experiments of Sir Isaac Newton. Could any means have been contrived to convince us more fully of the white appearance of light being the effect of a due mixture of all these colouring particles?

In order to your understanding the effects of the glasses, which are used for the purpose of astronomical observation, I shall explain all that is necessary for that purpose concerning the nature and properties of lenses.

A lens is a transparent substance of a different density from the surrounding medium, and terminates in two surfaces, either both spherical, or one spherical and the other plain, or both plain. A plain glass will refract the rays of light that fall obliquely on it, but will not collect them into a point.

A plano-convex is flat on one side and a portion of a sphere on the other, like A B, *Fig.* 3. *plate* 1. When parallel rays fall perpendicularly upon it, they pass through and are refracted, so as to meet in a point on the other side of the lens, called its focus; which spot is as far behind the lens, as is equal to the diameter of the sphere of which the convexity of the lens is an arc; as at C, which is at the extremity of the diameter of the sphere ABCDE. Convex lenses are called burning glasses, because the intensity of the matter of light from the sun, collected at the focus of a large convex lens, will burn or consume almost any substance fubjected to its influence.

If the rays of light are not stopped at the focus of a convex lens, as at C, but are suffered to proceed, they will, after crossing each other at the focus, diverge as from C to F and G, all but the middle ray H I, which, falling perpendicularly on the axis of the lens, will not deviate; and their divergence will be exactly equal to their convergence towards the focus. But if, when they have diverged to the same distance from the focus, as at K L, as they converged to it, they are not stopped, they will go on diverging continually, as from L to G, and from K to F.

If an object, as the Sun, be at an infinite distance, and its light pass through a convex lens, all the rays proceeding from it may be considered as parallel rays; and all these rays meeting in a point called its focus, the object will be transferred to that spot; so that by this property of convex lenses, objects the most remote are brought as it were close to our eye.

Suppose S, between M and N, to be the Sun at an immense distance, his rays will then appear parallel at the lens A B, as represented by the lines between M N and A B. After the rays have passed through the lens, they will meet at the focus C. If this lens be placed in an instrument properly constructed, the image of the sun will be transferred to the spot C, in its diminished state, and will then be brought near to the eye of a spectator, notwithstanding its great distance from our Earth: then by the aid of other glasses, called eye-glasses from their assisting the eye to view the object as near to it as possible, the angle under which the object is seen, may be larger than the angle the object itself would subtend in the same situation, as objects appear larger in proportion to the angles under which they are seen.

There are two kind of telescopes, the refracting and reflecting; the former do very well for terrestrial objects, but the latter are better adapted to celestial observations.

A reflector is furnished with a speculum, from which the image of the object is reflected to the eye, after being brought to a focus by the mirror; from which it passes to the eye by means of an eye glass, which refracts the pencils of rays, so that they are brought to their several foci by the humours of the eye; and according to their nearness to that organ, the greater does the object appear. The end of the telescope, which is always turned towards the object to be viewed, contains the object glass, and the other end contains the eye glass. According to the different convexities of these two glasses, the object is magnified; that is, its image, as seen by the eye through these glasses, is so much larger than the object itself appears to us without the aid of glasses. For a fuller illustration of this subject, I shall refer the curious to Dr. Hutton's Mathematical and Philosophical Dictionary.

Having contemplated one great agent, another offers itself to our consideration; not less important, but of whose essential principles we know still less than we do of Light. The effect of this agent is called Gravity, of which I have already mentioned some particulars, but shall now consider its general characteristics.

How important is this general principle by which the mighty fabric of the universe is cemented together, and yet its various elementary parts separated from each other! thus at once uniting and disjoining, its operations are carried on dependently and independently of each other.

When we consider this power retaining the planets within their orbits, and making them observe their proper distances from each other, and from the Sun; and causing them to perform their regular returns of periods, how do we admire the wise adjustment of this complicated yet simple machinery, and bless that Power who formed and directs this invisible agent; the absence of which, were it one moment to cease, or be diminished, would produce universal chaos.

As the great Newton could not define the cause of gravity, it is not likely that we shall be able to discover what that penetrating genius could not fathom; let us then be satisfied with the benefits we derive from it, and with the knowledge he has afforded us of its nature, without attempting to penetrate its essence :---" For who can by searching find out God."

The effects of gravity were never sufficiently considered till Newton, (to whom no effect in nature was barren of instruction) by comparing its effects in substances on our Earth, extended his speculations to every known part of the universe.

It is said that a very trifling circumstance first led Newton to reflect on the effects of gravity, in falling bodies in particular: sitting under a tree, an apple fell upon his head; we may suppose that the force with which it struck him, roused his attention, on account of the increase of pressure perceived from so trifling a substance to what it exerted when held in his hand.

The result of his meditations on this circumstance, terminated in inferring, that the force which falling bodies exerted, was in some measure independant of the body, and external; which principle being settled, he proceeded to investigate the effects of gravity in the most distant bodies in the universe; so that, by a very trifling circumstance indeed, the greatest discoveries were produced.

The Ancients were not acquainted with the universality of this property or force. They imagined that there were some substances which by nature were light, and therefore ascended; which opinion is now entirely disproved; as in what is called an exhausted receiver, all substances fall towards the Earth, and with equal velocity.

D

That gravity acts towards the centre of our Earth, is well known, by observations taken of the weight of bodies on different parts of the Earth's surface. At all parts of it which are equi-distant from its centre, their weight or force of gravity is equal; but as all places on the surface of the Earth are not at an equal distance from its centre, being more distant at the equatorial parts than at the poles, bodies weigh heavier at the poles than at the equator; and at every intermediate part of the Earth's surface, the weight has been found to be in proportion to its distance from the Earth's centre. This has been ascertained, by first observing the vibrations of pendulums of equal lengths, at London and under the equator, and then at different distances from both situations.

For the uses of the pendulum we are chiefly indebted to Galileo, who observing the vibrations of a lamp suspended from the roof of a house, discovered that they were performed sensibly in the same time, whether great or small; that their extent continually diminished, till the motion entirely ceased; and that the vibrations were slower in proportion to the lamp's distance from the point of suspension. This discovery delighted him exceedingly, as by it he perceived that he was furnished with a true measure of time.

As all circumstances which serve to illustrate the subject of motion are intimately connected with the Science of Astronomy, we shall contemplate a little the laws which govern the motions of pendulums, and the forces they are subject to.

Pendulums of the same length vibrate in the same time, whatever be the proportion of their weights. If one pendulum be four times as long as another, the shorter will vibrate in half the time, performing two vibrations in the same time that the longer performs one; so that it is the length of pendulums that determines the time of their oscillations.

A pendulum is a heavy body, suspended by an inflexible line from a fulcrum or prop, and is moveable round it; moving backwards and forwards to the same limits on each side of it. There is one point within every pendulum, into which, if all the matter that composes the pendulum were condensed, the time in which the vibrations were performed would not be altered by such condensation : this point is called the centre of oscillation. The point round which the pendulum vibrates is called its centre of motion.

Huygens was the first who demonstrated the properties of pendulums. He asserted, that if the point to which a pendulum is suspended were perfectly fixed, and all manner of friction and resistance removed, then a pendulum, once set in motion, would for ever continue to vibrate without diminution. But as this is not the case in practice, a pendulum which has received a certain impulse, will, unless that impulse be continually renewed, (as in the swing wheel of a clock) describe less and less arcs every time, till its motion is totally lost.

The motion of a pendulum being accelerated by gravity, like that of all falling bodies, proves, that in those parts of our globe where the same pendulum moves slower than it does at other parts of it, the force of gravity must be less in those parts ; and as gravity always draws towards the centre of our Earth, those parts must be further from that centre.

The time of each vibration of a pendulum is in proportion to the square root of its length ; that of ten inches in length making its vibration in half the time that one of forty inches makes its vibration in. Thus the fhorter the pendulum, the swifter it moves, according to the above ratio ; for as a pendulum oscillates always the same arc of a circle, the longer pendulum must be a longer time in describing the arc of its circle, because that circle is larger. The time in which the same pendulum performs its oscillations. depends upon its length, as well as upon the force which impels or draws it towards the Earth ; if the force be diminished, its motion will be retarded, and when this is the case, the pendulum rod must be shortened, to supply the diminution of force. By the application of the principles of motion in this instrument was the quantity of the increase of matter about the equator ascertained.

Sir Isaac Newton thus accounts for the increase of matter at the equator. The diurnal motion of our Earth being performed round an imaginary line which passes through it, the surface of our Earth at the equator, or part in the midst between the poles, must be farther from this centre of motion than any other circle parallel to it; and consequently those parts must move with greater velocity than any other, and have a greater tendency to fly off from the centre. All bodies in motion have a tendency to fly off from the centre round which they move, which is evident by a very simple experiment: Take a tassel of thread, and tie it to the end of a stick; then roll the stick quickly round in your hands, and the threads will endeavour to recede from the centre; and the quicker the motion, the greater will be the force they exert to fly off from the centre. Thus it is with the equatorial parts of our Earth, as the Earth is continually turning round a line within itself, called its axis.

As gravity or attraction acts on all parts of our Earth, and the centrifugal force is greatest at the equator, it is evident that that additional force I have been describing must, in some degree, diminish the power of gravity; therefore it is that bodies weigh less at the equator, not being impelled with so much force, on account of the reaction of the centripetal force. Mathematicians have compared these two forces, and have found the centripetal force to be about 290 times greater than the centrifugal, or that which is caused by the motion of the Earth on its axis; and that the Earth is of a figure resulting from the difference of those two forces; which idea has been confirmed by bodies weighing only 289 pounds at the equator, which weigh 290 near the poles.

The scientific Newton instructed mankind in those laws, by which these effects in nature are now ascertained, although the great efficient cause of them cannot be understood. His assertions have been proved by subsequent inquiries, and are established by experience, as will be seen when we come to treat of the mensuration of the Earth.

Thus much I thought it necessary to premise, before entering upon the more immediate subjects of Astronomy; as without these leading principles of this stupendous Science, separately and previously impressed on our minds, we might have found the aggregate too mighty for us, including too many ideas to be digested together, and implying more than could well be comprehended by young minds. Ah! what is there in Nature which does not, with all our pains, with all our scrutiny, contain much more than we can comprehend? Nothing is perfectly revealed to us. What is light? Who can investigate its essence, or determine its essential properties? Yet much has been ascertained of it by the incomparable Newton, whose well-conducted experiments and conjectures, founded on similar effects perceived of things we do understand, are inferences not to be rejected, being the result of sound reasoning and steady application. Such are all the truths advanced, and hypotheses framed by this excellent man; his researches were not made in a detached or desultory way, but each new subject referred to what went before; and no wandering ideas had room to enter, so closely was the firm texture of his philosophical inquiries woven together.

Having advanced sufficiently on the subjects of light and gravity for our present purpose, I shall not extend them farther at this time, but beg leave only to conclude them with reflections which the sense I entertain of the weakness of human reason, when employed to understand the energies of the Divine Mind, aptly prompt me to make. Of nothing can we speak with certainty; we cannot account for all the operations of those senses which we continually perceive the effects of, or comprehend in what manner they make us perceive the effects of causes more remote. Who can say how the organs of vision convey to our minds the idea of visible objects? What is our mind, our understanding? who can explain those things?—No mortal can. Then why, when we acknowledge ourselves so totally ignorant of our own nature, why should we expect to understand the sublimer energies of divine truths, or to comprehend the nature of spiritual existences; or why with-hold our assent to evidences which, when we compare them with the necessities of our nature, we find as fitting to them as the organs of our bodies to their functions.

END OF THE FIRST LECTURE.

Plate II.

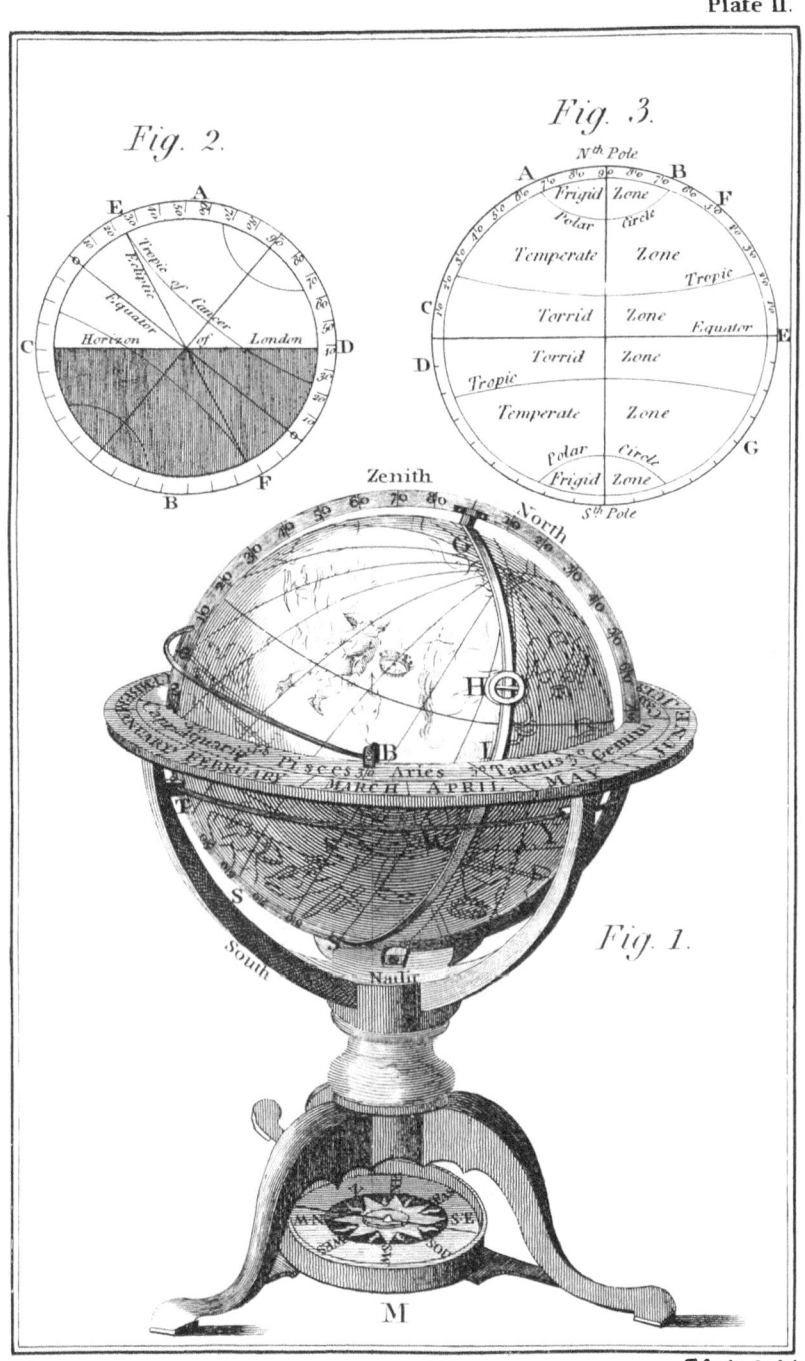

Fig. 2.

Fig. 3.

Fig. 1.

Zenith

North

South

Nadir

M

T. Conder Sculp.ᵗ

Plate III.

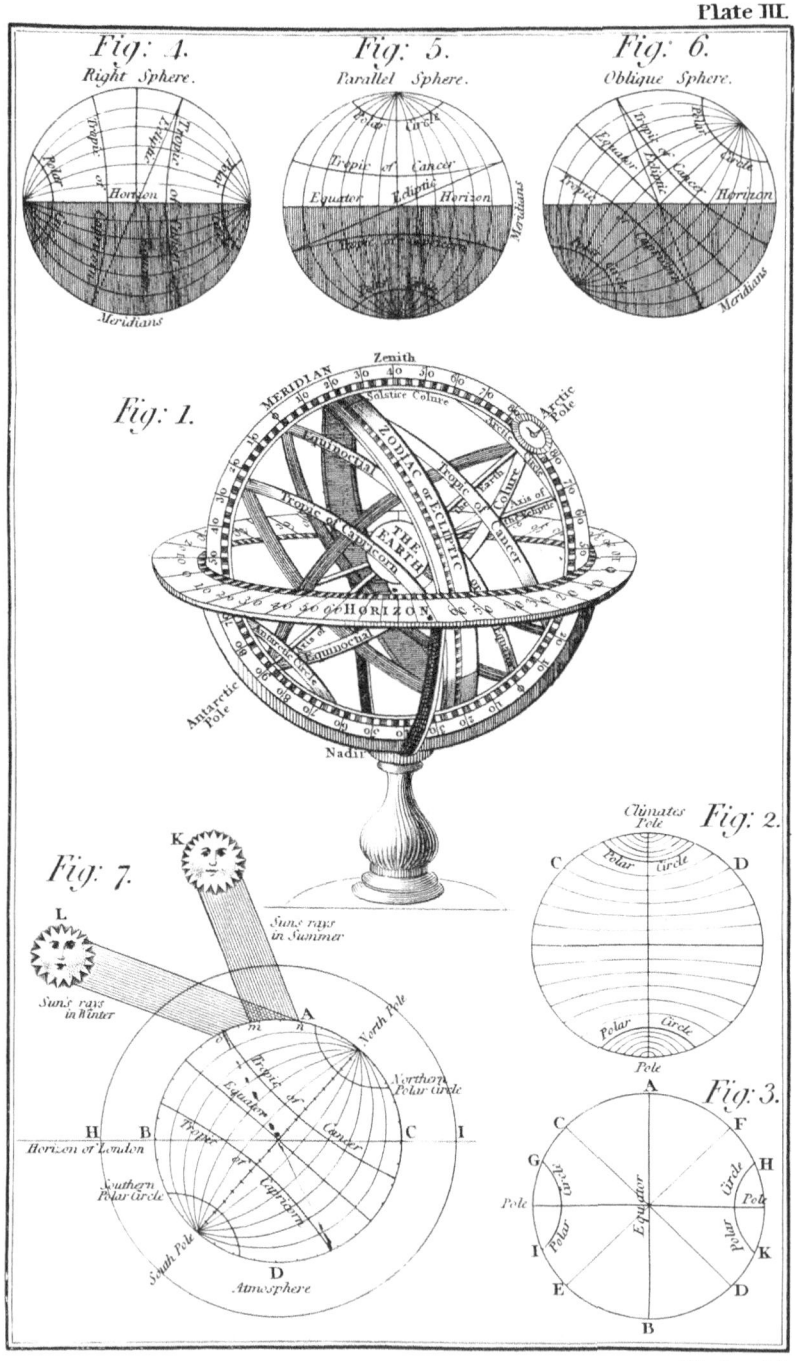

Fig: 4.
Right Sphere.

Fig: 5.
Parallel Sphere.

Fig: 6.
Oblique Sphere.

Fig: 1.

Fig: 2.

Fig: 7.

Fig: 3.

T. Conder sculp.t

LECTURE II.

ORIGIN OF ASTRONOMY.——ITS FIRST CULTIVATORS.——CHALDE-
ANS, ÆGYPTIANS.——FIRST MEASURE OF TIMES AND SEASONS.
——ORIGIN AND SYMBOLICAL EXPLANATION OF THE ZO-
DIAC.—— CONSTELLATIONS ; THEIR USE IN NAVIGA-
TION.—— ORIGIN AND PROGRESS OF THIS ART.——
HISTORY OF ASTRONOMY.——THE CALENDAR.

THE descendants of Noah, by their numerous progeny, being obliged to extend themselves for the purpose of supplying their families with necessary food, formed a numerous nation in the wilderness of Shinar ; where the cultivation of the land becoming their principal object, it was natural for them to endeavour to ascertain the return of particular seasons, in order to provide for the exigencies of each.

The human mind has always been found capable of providing for the emergencies of our present ſtate, when called into action, although by the feeble efforts observable in some minds, one would be led to consider it as a poor resource ; however, this imbecility must not be accounted the natural, but rather the infirm and accidental state of the mind, produced by inactivity.

No doubt, when the importance of the observation first caused the Chaldeans to reflect on the motions of the heavenly bodies, they were surprized that curiosity had not previously led them to observe, what necessity then compelled them to investigate.

The Chaldeans saw, that at some periods the Sun afforded them more of his influence than at others ; and perceived, that according to these circumstances their corn in ripening was either benefitted or otherwise, so as to produce full ears or scanty and imperfect grain. But how were they to enjoy the benefit of the former effect, or to avoid the disadvantage of the latter, unless they could foresee the returns of these periods? To obtain the desirable purpose of knowing when particular seasons would return, the Chaldeans found the courses of the Moon particularly well calculated: perceiving her to rise and set to their horizon, to change her places of rising and setting each time, and to vary her form, they took notice of the time in which she performed all her changes; and they observed that after one course was accomplished, all the same circumstances were renewed and repeated as before. By remarking how often the Moon performed her changes from one state of the Sun, in respect to that part of the globe, to another, they furnished themselves with regular periods, and were thus enabled to avail themselves of the advantages offered them by each season.

Deriving such great advantages from the Moon, it would not have been surprizing if they had paid that adoration to her which is ascribed to them by some historians, although apparently without just grounds for such an assertion. Why should their festivals, at the time of the new Moon, be addressed to her as their object of adoration? It is more rational to suppose it as the period fixed upon to offer their public thanks to the Deity for so great a gift; and the new Moons, which renewed the seasons, being kept the most solemn of all, serves to strengthen this latter opinion, as it proves that the benefit derived from the Moon, by the information it conveyed, was what excited their praise and thankfulness.

As they had not at this period classed the stars in constellations, they could not note the progress of the Moon by their aid, but only by her different appearances and situation in respect to the horizon, to observe which they assembled on high places or in deserts.

Their periodical sacrifices always ended in a repast, at which, what had been sacrificed was eaten with gratitude by the whole company. These devotional and convivial meetings tended, no doubt, to harmonize their minds and conciliate their mutual good-will.

The festival of the new Moon continued for a considerable time, and was observed in many nations; we read of its being a custom among the Hebrews, the Egyptians, the Persians, the Turks, the Greeks, and the Romans, down to the time of the Gauls.

After the restitution of the observation of times and seasons by the Moon, the next thing which appears to have excited the attention of Astronomers was the dividing time into years; for which purpose they began to group the stars, that they observed the Sun to pass by in his apparent revolution, into constellations, under some familiar figure emblematical of the respective seasons in which the Sun was in those signs.

This improvement is ascribed by some authors to the Egyptians, by others to the Chaldeans, which latter I have ventured to adopt, as the arguments in favour of that decision I think incontrovertible; which are, that admitting the character ascribed to each constellation in the Zodiac to be symbolical of the seasons, as no doubt they were, they, having no other types of their ideas but what were expressed in this hieroglyphical manner, the application of them to the seasons in which the Sun passed through those signs answered exactly to their Shinar, but by no means to Egypt, which at the season represented by Virgo, or the Gleaner, is inundated by the overflowing of the Nile. The waters not withdrawing till the latter end of October, the time of harvest in Egypt is not till March and April, as the grain cannot be sown till November.

This circumstance, doubtless, entirely invalidates the idea of the Egyptians being the dividers of the Zodiac; and to suppose that names were given to those constellations, and that they should include such a certain portion of that circle in the heavens without its being previously divided, is an incongruous idea, and not to be

E

admitted : therefore we may allow, with many others, the division of the Zodiac to be more ancient than the Egyptian colony. Have we not good reason then to ascribe it to the Chaldeans? They being the first people after the flood, and being compelled to investigate some of the celestial phœnomena, it is most probable that after they had advanced so far as to calculate the periods of the Moon, they did not stop there; but that finding the advantages resulting from the consideration of one of the heavenly bodies, they prosecuted their researches, in order to ascertain that period in which all the circumstances of the seasons had been passed through. And how could they effect this but by distinguishing the situation of the Sun in respect of the horizon and particular stars?

Another argument in favour of the Chaldeans being the dividers of the Zodiac, offers itself. They never could have ascertained the returns of the seasons by the observation of the Moon alone; as, although in twelve revolutions of that body, the Sun would have nearly performed his apparent revolution, yet not entirely so; therefore, if they had been guided by observation of the Moon's period alone, compared with the annual period of the Sun, they would have made great mistakes, and could never have calculated the returns of the seasons with accuracy.

The method said to have been pursued by these astronomers, in the infancy of science, in order to ascertain the important epocha in which all the circumstances of the seasons had been passed through, and afterwards of dividing the Zodiac into twelve parts, was as follows: They prepared a couple of vessels; piercing a hole in the bottom of one of them, they placed it over the other vessel : after putting a plug into the orifice, they filled the upper vessel with water, and left them in that situation. Observing the time a particular star, in that circle of stars which the Sun appeared to pass through, transited the horizon, they instantly withdrew the plug, and suffered the water to run into the lower vessel, which it did gradually and almost regularly; letting the operation continue till the same star again passed the horizon the next evening, when withdrawing the under vessel, they provided themselves with the measure of time of one intire revolution of the heavens.

The way this ingenious people prosecuted their researches on this subject, was by procuring two vessels, each of which would contain one-twelfth part of the whole quantity of water employed in the foregoing operation. Placing one of these vessels under the large one with the hole in it, they poured the whole quantity of water before employed into that vessel, and waited for the appearance of that remarkable star at the horizon which they had before started from;—when they withdrew the plug, and let the water run into the small vessel; which, when full, they removed, and placed the other of the same size in its place, letting it be filled in the same way; and thus alternately they filled the vessels three times each, observing at the beginning and end of each operation, what particular star was at the horizon. Thus did they perform six divisions of the Zodiac;—for the other six, they were obliged to wait till a proper season for the observation offered.

Those acquainted with the science of Hydrostatics and Hydraulics, or to whom I have had the pleasure of delivering my lectures on those subjects—know, that the velocity with which water flows from a given aperture, is in a certain proportion to the height of the fluid. Whether the Chaldeans were acquainted with this circumstance I know not.—To the first part of the knowledge to be derived from the operation alluded to, it certainly was not essential, but to the subsequent part, in the way they performed it, it undoubtedly was;—from which we may infer, even supposing them to have been unacquainted with the law before mentioned, they must have observed the different velocity with which the water flowed, and consequently, that they made some allowance for it; otherwise, their divisions and subdivisions could

not have been so exact :—and admitting the necessity of this observation, it is most probable that their manner of conducting this
experiment, although in some respects properly described, may
not be so in every part of it.

I imagine that the notion of the Egyptians being the dividers of
the Zodiac, might originate in the symbolical writing continuing,
and being much extended in their time; as they certainly added
other figures, adapted to the circumstances of their peculiar situation.

The overflowing of the Nile being to them an epoch of great
importance, they were led to make observations on those stars,
which, when in a certain situation in respect to their horizon, fore-
told the approach of that inundation.—The star which by its
magnitude presented itself most particularly, was that which they
afterwards called the Dog-star, from the affinity the service it rendered them had to that which the house dog affords his master, by
warning him of approaching danger :—this star always appearing
in the morning before dawn, immediately preceding the overflowing of the Nile;—on discovering which, when in a certain situation, they retired to the higher lands, first providing themselves
with every thing necessary for their temporary retreat.

They included that remarkable star under the figure of a dog's
head, to which they affixed a human body, with a kettle on its
arm, and wings on its feet,—thus describing the circumstances of
their temporary retreat; which continued whilst the Sun was in the
signs Leo and Virgo; to express which, they placed on the edges
of terraces the figure of a sphinx, made up of a female face and
the body of a lion couching.

Having spoken of the overflowing of the Nile, it may not be
unpleasing to thofe who are not acquainted with the cause of that
inundation, to inform them to what it is ascribed.

The Nile flows through the middle of Egypt, and falls into the Mediterranean sea by seven outlets. The overflowing of its banks is owing to the heavy rains which regularly fall in Ethiopia (which is situated south of Egypt) in the months of April and May; the waters of which, rushing down upon the low country of Egypt, lay great part of it under water; which, although it occasions some inconvenience to the inhabitants, yet repays them by the fertility it produces. And thus may we perceive, in all the dispensations of Providence—temporary evils more than alleviated by permanent good.

In referring to that part of the symbolical writing of the ancients, which is connected with the science of Astronomy, we may first examine the inferences deducible from the figures of the Zodiac, in respect to the seasons, as handed down to us by historians.

The Spring is distinguished by Aries or the Lamb, Taurus or the Bull, and Gemini or the Twins; which latter were formerly represented by Kids; and as these animals are produced at this season, it is supposed that circumstance is meant to be thus expressed.

The first month of the next quarter, being represented by the Crab, admits but of one solution — their having perceived that after the Sun had arrived at a certain height above the horizon, he returned back again declining from that height, they meant to express this retrogression of that luminary by an animal which walks backwards.

When the Sun was passing under the next constellation they felt extreme heat, and therefore they introduced the figure of a Lion, denoting the raging sensations at that time.

The character given to the next constellation seems equally appropriate, denoting that season in which the gleaner gathers the refuse of the harvest fields.

The equality of the days and nights when the Sun enters Libra, is naturally represented by the Balance; as are also diseases which follow the great heat, by the Scorpion.

The next constellation is symbolical of the fporting season. The next is characterized by a Goat, that animal, climbing rocks, being supposed emblematical of the Sun again ascending towards its highest situation.

The circumstances of the rainy season are expressed in the figures of Aquarius and Pisces.

The circle which includes these figures is called the Zodiac, because it is a circle of animals.

Thus did men, by a little use of their reason, fix a standard for regulating all the affairs of life. How rich a gift is reason! and how all-sufficient for our necessities, when properly exercised; but how base, as well as inconvenient, is the abuse of it.

Symbolical writing was most probably produced by Astronomy, and was no doubt found greatly useful in acquainting people with events that would happen at certain seasons. Yet the multiplicity of figures necessary to express all ideas by this mode, must have rendered it very irksome; which caused man's invention to construct other characters, which, by their different combinations, might serve to express all the different emotions of the mind, and record all circumstances; and when this great object was accomplished, it became the popular mode, and symbols were totally laid aside.

Symbolical writing was productive of one great evil to after ages; being misconceived, it produced that hydra mischief called idolatry:—These symbols remaining in places of public worship, where they had been placed in reference to those seasons when festivals were held, their signification not being known, false ideas were annexed to them, by which the people became gross idolators.

It is recorded that the Egyptians, having placed a symbolical representation of the Sun, under the figure of a man, with a scepter in one hand and a whip in the other, denoting his rule over the day, and his regulating the course of nature; succeeding ages supposed that this figure represented the man who ruled the heavens; that he lived in the Sun, having been translated from earth to heaven; that he watched over Egypt, and had a particular affection for it;—they called this suppositious deity God, Jove, Ammon.

Thus not being able to comprehend the proper attributes of these figures, they invented circumstances to account for the attributes themselves; and this we may naturally suppose was the origin of idolatry.

The feasts of the Chaldeans and Egyptians, thus perverted, were communicated to the neighbouring nations, and proved the origin of fables, metaphors, and a variety of deities, which produced that mixture of truth and falsehood found in the fabulous history of those times.

It is also imagined that the figures of the Zodiac gave rise to the worship of animals, and to the doctrine of transmigration.

Omens were also much used in those times, such as the flight of birds; which ridiculous custom originated in those figures which were intended to represent the winds. In short, the figures of men and animals, contrary to their original signification, were converted into celestial and infernal powers, and the whole world was filled with superstition and idolatry.

But let us now leave this barren retrospect, and return to the progress of astronomical knowledge, and to the uses to which it was applied in ancient times.

As the people became divided into different nations, the exigencies of trade and navigation led them to consider the stars more attentively; and they observed some stars which never set below the horizon, and which were seen constantly very nearly in the same place; these served them, though not accurately, as a guide to steer their course by.

The group of stars thus distinguishable by their brilliancy and peculiar situation, was that we now call the Great Bear, but which the Phœnician pilots called Parrasis, which signified the Rule, the Guide, and sometimes Doube, signifying that which gives advice, and also in the Phœnician language the Bear. In this latter sense it was not intended to be understood; yet being received according to that idea by the Greeks, they represented it under the figure of a bear, which was quite foreign to the meaning, though it remains the same to this day.

The Phœnician navigators finding that this constellation, now called Ursa Major, or the Great Bear, occupied a large space, and that it made a very extensive revolution, which subjected them to considerable deviations from the true course, they sought for some constellation less variable in the situation of its parts; when they discovered one of the same form but of less dimensions, and nearer to the pole, and of course less liable to change of situation : this was the constellation we now call Ursa Minor, or the Little Bear.

One star in this constellation, by its situation, which is very near the axis of the apparent motion of the heavens, served them for an accurate guide, when it was not obscured by clouds : by this discovery, navigation was much benefited and extended.

The observation of the polar star afforded the Phœnicians the opportunity of becoming a wealthy and respectable people. Their territory, although but a border of Syria, became opulent, in consequence of their extensive commerce to all the coasts of the Mediterranean : they possessed themselves of very useful ports in Spain, especially the modern Andalusia, which at that time went by the name of Tarsis; the fruitfulness of the vines, the vast quantity of timber, corn and cattle, and the mines of gold, silver, and tin, made this country the principal object of their attention.

On account of the extent of this voyage, they were obliged to build large ships; from which circumstance, all ships of large dimensions were called ships of Tarsish. Possessing an enterprizing disposition, they were not content with this boundary, but after a time they ventured to pass the Streights; which undertaking, we are informed, proved very advantageous to them, as by that means they procured a safe asylum for all the riches they brought from other countries, in the island of Cadiz, which was then called Gadir; other nations not being sufficiently skilled in the art of navigation, to approach them in that harbour.

The success of this enterprize made them bold in forming new ones, which they prosecuted with the same ardour, and with the like success; so that finally they insured to themselves an advantageous commerce, not only to all the coasts of the Mediterranean, but also to Asia and Africa, through the gulph of the Red Sea.—Thus, by the observation of only one star, the Phœnicians became the richest people in the world; and they appear to have been the first who extended navigation to any considerable length.

This science was communicated to the Hebrews by the pilots of Hiram, king of Tyre, about 1000 years before Christ, which was before the time that the Greeks had any knowledge of it.

History acquaints us, that the cause of the prosperous and happy state of the Tyrians did not escape the attention of the wise prince Solomon, who perceiving that navigation was the parent of opulence, and an extensive marine establishment the dispeller of idleness, obtained the service of the pilots of king Hiram to conduct his fleets, which he had established in Elah and Ezien-geber: That the Tyrians and Hebrews went together to Ophir, and brought from thence immense quantities of gold, jewels, valuable woods, and ivory; and that they were three years in performing this voyage.

A considerable trade is at this day carried on with Africa, where this Ophir was, for gold-dust, which the torrents of rain scatter on the coasts, after having washed it from the mines, which are very

F

numerous in that country, especially in the mountains of Manica, from whence descends the river Saphara.

These voyages were discontinued about two thousand years, after which the Portuguese renewed them, about three centuries ago; and it is said, that they ascribed to themselves the first attempt at this navigation.

The commerce which was carried on so long by the Phœnicians and Hebrews, was destroyed by the kings of Babylon; yet although their long voyages to Cadiz were interrupted by the conquests of these ambitious monarchs, they did not lose the recollection of them.

Herodotus informs us, that Neaco, king of Egypt, who reigned 600 years before Christ, wishing to restore the ancient fplendor of that kingdom, found he could not succeed without navigation, and therefore appointed an extensive marine establishment.—Among other projects, it is said that he attempted a junction of the Mediterranean sea with the ocean, but was obliged to abandon that project.

The Phœnicians, although not at that time in the practice, yet were well informed in the theory of the navigation of the Red Sea, by the accounts handed down to them from their ancestors; they therefore readily undertook to conduct the fleets of Neaco along the coasts of Africa, in their voyage to the parts through the Red Sea.

It was Thales the Milesian, who about this time (six hundred years before Christ) taught the Greeks of Ionia the use of the pole star. Travelling into Egypt, Crete, and Phenice, in search of knowledge, he became acquainted with the advantages resulting from the observations of this star, and communicated them to the Greeks of Ionia at his return, from whence, in process of time, the knowledge of the stars was conveyed to all parts of Greece.

Before this communication of the use of the polar star, the navigation of the Greeks was very circumscribed, as may be inferred from the manner in which the Ancients celebrated the expedition of the Arganauts, which was made across the Propontis, now the Sea called Marmora, on account of the great distance they ventured from shore, though it is only three hundred miles in circumference, lying between the Streights of the Dardanelles and that of Constantinople.

To express their admiration of this bold enterprize, as it was thought to be at that period, they made a constellation of the ship which had passed from Colchis to the mouth of the Phase. Thus we perceive the difficulty of navigation before the siege of Troy.

Virgil has expressed the timidity of the ancient navigators, by making his Grecian hero coast along from place to place, agreeably to that propriety which characterises all his descriptions, which in date and locality perfectly accord with the time and circumstance he relates, and, in this instance, is the more remarkable, because in his time navigation had been much improved, and therefore it required the greater degree of judgment to keep exactly to those circumstances which preceded it.

Those who are not acquainted with the expedition of the Arganauts, previously referred to, will find a full account of it in Lempriere's Bibliotheca Classica, or Classical Dictionary; and a map of the country I have been adverting to, with both the ancient and modern names of places, &c. in D'Anville's Atlas, or Body of Ancient Geography.

Thales was justly called one of the wise men of Greece, as his extensive knowledge in geometry, astronomy, and philosophy, testifies.

It was this great man who discovered the solstices and equinoxes, and divided the sphere into five zones. He also divided the year into 365 days.　He died at the age of 95, about 548 years before Christ.

In Thales we have an example of the utility of observation, judiciously taken, and how a mind prone to study, rises progressively from one truth to another, and the extension the mental power is capable of, which ought to excite in us the like application in all we attempt to investigate.

I love to trace the gradual progressive elevation of ideas, from their first dawn, in those minds that have indicated such a sublimity of sentiment and clearness of intellect as are not common; I cannot help participating in the satisfaction they must have felt, who by a due cultivation of their understanding, have rendered their fellow creatures such essential services,—and I honor them from my heart.

The manner in which the great philosopher Thales determined those phœnomina, the discovery of which is ascribed to him, was by due observation of the stars under which the Sun passed in its apparent revolution, and those passed by the Moon.　He perceived that they did not both pass exactly under the same, but that the orbit of the Moon cut that of the Sun in two points, deviating a little from it on each side, and that it did not always cut the orbit of the Sun exactly in the same place each month, but that these variations returned nearly in the same order after a certain number of revolutions of the Moon.

Whether this great man was sensible of the benevolent design of our Creator in this arrangement of the nodes of the Moon, is not said, but it seems an impeachment of his judgment, derogatory to his character, to suppose him ignorant of the advantages accruing to all the inhabitants of the Earth from this wise ordination of Providence, who, but for it, would, once a fortnight, be deprived of necessary illumination.

As we are certain that he understood the consequence of their being in a direct line, by his calculating the time of the eclipses, it must have been impossible for the utility arising from the deviations from a right line to have escaped his observation.

This discovery of the cause of eclipses relieved the people from the dreadful apprehensions formerly attendant on those temporary deprivations of the Sun's light, which they had supposed to be occasioned by the wrath of God.

How ignorant the Medes and Lydians were of the cause of these phœnomina, is evident from the effect a total eclipse of the Sun had on them in the time of a battle, in the heat of which the Sun being suddenly obscured to the place at which they fought, they desisted from all hostilities, and all animosity subsiding on the evidence, as they thought, of divine indignation, in which they were mutually involved, as recorded by Herodotus.

Geography was also much benefited by this discovery, particularly by the observation of lunar eclipses, which served to give juster ideas of the distances of countries.

Thus, suppose two spectators placed at different parts of the globe, and each prepared to notice the instant of the Moon's entering into the shadow of our Earth; also the time of its total obscuration, and of its emersion. Then on comparing the two times noted by each observer, they were enabled to judge of the comparative distances of the places from each other, east and west, even supposing them only acquainted with the apparent motion of the Heavens. Lunar eclipses also served to ascertain the rotundity of the Earth.

Before this time, the Earth was often supposed to be a flat surface, surrounded with water; which opinion was strengthened by the writings of the poets, who always expressed the Sun's rising above the ocean, and setting in it to cool himself.

This vulgar prejudice was shaken off by those of the Ionian school, who becoming acquainted with the circumstances of the

Moon shining by borrowed light, perceived by the form of the Earth's shadow projected on it, that it must be globular.

The Ionians seem to have been an ingenious and sagacious people, and to have possessed great steadiness of character; probably the salubrity of the climate might contribute to these advantages, as we know that soundness of constitution is generally accompanied by strength of understanding, and perseverance in all the mind seeks after.

Thales was so happy as to find in his disciple Anaximander the same disposition for knowledge, and ardent spirit of enquiry, as he himself possessed; so that the congeniality of their tempers made them firm friends and allies.

How great was the advantage of thus being enabled to compare their ideas together! for however penetrating a genius individuals may possess, their advancement in knowledge will be much retarded without the aid of another of similar talents; as it is by comparing their opinions, that they can establish them with facility, and assert them with confidence. If the understanding does not rise above mediocrity, I can assert from experience the disadvantage of not having a second person to consult on its opinions; as I have felt the impediment it has been to the prosecution of my present study, by being obliged to digest the opinions of all writers on the subjects of these Lectures without any assisting friend to strengthen or confute my ideas as they arose.

Although Astronomy was much advanced in the time of Thales the Milesian, yet Geography must have been very imperfect, as it was not till the time of Newton and Huygens that the true figure of the Earth was discovered.

Before their time, all the calculations of the distances of places, &c. were made on the supposition that the Earth was a perfect sphere; but these great men, by the known laws of gravitation, discovered that the true figure of the Earth was that of an oblate spheriod.

Notwithstanding the imperfection of geography in the time of Thales, it was sufficient to enable the scientific Greeks to enlarge their territories, and keep the command of the ocean, in spite of the formidable fleets of Persia, whose empire they at last destroyed; and for this victory over a powerful enemy, it is said they were indebted to their extensive learning, not to their forces, which were by no means equal to those of the Asiatic monarch.

After the death of Thales, Pythagoras taught the Europeans that the Earth and Planets turned round the Sun, which luminary was, in respect to them, at rest in the centre of the system; and therefore that the diurnal motion of the Sun and Stars was not real, but occasioned by the motion of our Earth on its axis in a contrary direction; and this idea is ascribed wholly to him. He studied astronomy in Egypt, where he went for that purpose, and he remained there a considerable time, consulting with the priests and Magi on all the subjects of Astronomy. The instruction they afforded him, he communicated to the Grecians at his return, together with his own improvements of that science.

After Pythagoras, astronomy was neglected for about two hundred years; most of the celestial observations brought from Babylon, we are informed, were lost, few only being revived by the Ptolemys, who renewed the study, and sought eagerly after the works of the ancient astronomers, thinking no knowledge comparable to that of the system of the world; by which means Alexandria became the school of that science.

The ideas they formed were totally different from those of Pythagoras, for which their judgment ought not to be impeached, as most probably they were not possessed of any vestiges of that more natural system; because after such a lapse of time since the science had been cultivated, many remnants of it could not be expected, at least of the speculative part.

Eratosthenes, keeper of the library of Alexandria in the time of

Ptolemy Evergetes, undertook to calculate the number of ftades *
that might make up the circumferonce of our Earth; and as he arrived somewhat near the truth, even in the infancy of this science,
it may not be thought improper or intrusive, if I describe the method by which he ascertained it: and I shall have pleasure in acquainting you with the ingenuity of the mode, as I always feel
great delight in communicating circumstances expressive of the sagacity of that active principle, the human understanding.

Eratosthenes observed, that in the summer solstice the Sun passed
through the vertical point of the city of Sienna, which was situated
on the confines of Egypt and Ethiopia, under the tropic of Cancer. In the situation to which the Sun would be vertical on that
day, he had a well dug, which at twelve o'clock on the day of the
summer solstice would be wholly enlightened within side by that
luminary. At the distance of 150 ftades round this well, posts
raised perpendicularly cast no shadows. He had a concave hemisphere made, which he placed in the city of Alexandria; raising a
post perpendicularly in the centre of it, the top of this post or
needle was exactly at the centre of a sphere of which that was the
radius. Remarking on the appointed day, at the instant of noon,
the distance of the Sun from the vertical point, by the shadow of
this post, he found at that time, on that day, at which the Sun
was vertical to the city of Sienna, it wanted one-fiftieth part of
the whole circle to be so at Alexandria;—this he discovered by
the distance of the shadow of the top of the post from the foot of
the needle; from which observation he naturally inferred, that the
city of Alexandria must be one-fiftieth part of the whole circumference of the Earth distant from Sienna.

The distance between those two cities was easily measured; and
when ascertained, he multiplied it by 50, and thus discovered that

* A stade contains 925 paces, a pace 5 feet.

the Earth was 250,000 stades in circumference; the distance between Sienna and Alexandria being 5000 stades. Although this was not an accurate measurement, yet it was as nearly so as could be made by those unacquainted with the laws of gravity, or the more elevated principles of the mathematics.

Hipparchus, who lived at this time, distinguished 1022 stars, by giving names to them, and placing them in their true situations in respect to latitude and longitude.

Succeeding Astronomers have added considerably to the catalogue; Dr. Halley, Mr. Flamsteed, and others, having augmented the number to near 5000, and ascertained their situations with the greatest accuracy.

Although the stars have each a name, thofe names are for the most part expressed only by the letters of the Greek alphabet;— those which are the most conspicuous in each constellation being marked with the first letter, α; the next with β, the second letter in that alphabet, and so on; yet there is one exception to this rule, which is, that any stars which require particular notice, although they are not the most conspicuous in the constellations to which they belong, are distinguished by that character used to denote one of the first class.

We are told that our ancestors, at least the Druids, were not inattentive to the phœnomena of the heavens; but the Romans were by no means a scientific people, military discipline and politics seem to have engaged their whole attention, while physics and astronomy were overlooked by them; and that necessity, rather than inclination, (as we may naturally suppose from the character of the people) rendered Scipio, Pompey, and Julius Cæsar favourers of this admirable science.

Always occupied in ambitious projects, which could not be effected without the aid of navigation, it became necessary to them to be acquainted with times and seasons, places and distances.

G

Julius Cæsar, by his perpetual and far extended conquests, became one of the most eminent geographers of his time; this he could not have been without some attention to Astronomy; we therefore naturally conclude that notwithstanding his active and enterprising spirit led him to martial exploits, yet he sometimes relaxed himself in the harmonizing and calm delights of astronomical observations; and that he did so is evident from his manner of addressing one of the priests, on his reformation of the Calendar, viz. "I expect that Eudoxus will not in future be more famous for the Ephemerides which he gave to Greece, on his return from Egypt, than I shall be by the order to which I have reduced the course of the year." Which prediction was verified.

I do not think that it could afford you either entertainment or useful instruction were I to relate all the different subdivisions of time, or variations in the Calendar, since the course of the year was established; I shall therefore only mention such circumstances of it as may serve the purpose of necessary information, and justify my assertions respecting it.

The month called July derives its name from Julius Cæsar, he having rendered the mode of computing the year more conformable to the apparent annual course of the Sun; which caused his name to be given to the most delightful month in the year, in compliment to his abilities. Although he reformed the Calendar very considerably from what it was in the time of the Greeks, yet as his computation was made on the supposition that the Sun performed his apparent annual revolution in exactly 365 days and 6 hours; which is contrary to the fact, as it is performed in 365 days, 5 hours, 48 minutes, and 49 seconds; therefore, according to his computation, the civil year must have exceeded the solar by 11 minutes 11 seconds, which in 130 years amounted to one revolution of the Earth on its axis; a space of time that altered the times of the Sun's situation one whole day; which, in the course of

47,450 years, would have entirely changed the seasons, from the time they were first calculated for.

Notwithstanding this incorrectness in the Julian Calendar it was adopted by all the states of Europe, being considered at that period a perfect computation of time.

Julius Cæsar's reformation of the Calendar arose from his having imagined that the Sun completed his apparent revolution in 365 days 6 hours, whereas amongst the Greeks it was computed to be performed in 354 days, which must have rendered their Calendar very imperfect and embarrassing.

In order to allow for the odd 6 hours in each year, Julius Cæsar introduced an additional day every every fourth year, and omitted the 6 hours in the three intervening years, making each of them to consist of only 365 days. He made the first Julian year to consist of 444 days, which caused great confusion; his motive for this was to allow for the 90 days which had been lost by the former mode of computation. In order that the courses of the Moon might agree with the seasons in his Calendar, he divided some of the months into a greater number of days than others.

The first who began to discover the imperfection of the Julian Calendar were Bede, Sacro Bosco, and Roger Bacon; perceiving that the true equinox preceded the civil, they set about rectifying this error; and having found the cause of it, they calculated what must have been the difference from the time that the vernal equinox was fixed by the council of Nice, in the year 325, and they found that it had differed from that time to the year 1582, about 10 days.

Although these great men discovered this error, yet it was not reformed by them, this being effected by Aloisius Luilius. As in the very year those had made the discovery, Pope Gregory XIII. had the old Calendar abrogated and the new one established, under the appellation of the Gregorian account, or new stile, which is the one now in use in most parts of Europe.

The first step that was taken to render the circumstances attending the Sun's apparent motion conformable to calculation, was lopping off those ten days which had been gained, and which had displaced the equinoxes, so as to bring them to correspond nearly with the 21st of March, and the 22d of September. A similar alteration in the Calendar was made by the parliament of England in the year 1752, when the difference had amounted to eleven days, which eleven days were taken from the month of September; and then, by calling the 3d of September the 14th, they brought the autumnal equinox to the proper place, which it was in at the time the council was held in 1752. In order to retain the equinoxes in their proper situation, it was ordered, that three days every 400 years should be omitted in the following manner: The years 1800 and 1900, which should have been leap years, were to be computed as common years, containing only 365 days each; that the year 2000, and every fourth hundred year after that, should be a leap year, containing 366 days, the intermediate hundreds being only common years. And by this judicious arrangement, our reckoning will not vary a whole day from the true time in less than eight or ten thousand years; a difference so small, as not to be at all material.

Our times and seasons now correspond with those settled by the first Christian council, in the time of Constantine the Great, when the festivals of the church were fixed by his order, in the year of our Lord 325.

Having explained the calendar sufficiently for my purpose, those who wish for a further elucidation of the subject, or mathematical definition of it, I beg leave to refer to that useful oracle, the Mathematical and Philosophical Dictionary of Dr. Charles Hutton.

The first seven letters of the alphabet (A, B, C, D, E, F, G,) are set to the days of every week, and repeated over and over again from the beginning to the end of the year, viz. A to the 1st day of the year, B to the 2d, C to the 3d, and so on till G on the 7th; then, over again, A to the 8th day, B to the 9th, &c. So that the same letter falls upon the same day of every week in the year; and the letter which falls on the first Sunday, and every other Sunday after, in the same year, is called the Dominical or the Sunday letter for that year. But as the 365 days of an ordinary year contain one day over the exact 52 weeks, the Sunday letters will fall back one place every year; so that if the Sunday letter be G for some year, it will be F the year after that, and E the second year after, &c.

As the intercalary day introduced into the calendar by Julius Cæsar, and which still continues in use, being allowed for in February of the leap year, might otherwise have caused some confusion, these first seven letters of the alphabet are used in the following manner: the 28th and 29th of February in the Bissextile have but one letter assigned them, so that the following Sunday goes back a letter, and so on for the rest of the year. As thus—

Suppose the dominical letter in leap-year to be C; then after the 29th of February, the Sunday letter will be B; and if in leap-year the 1st of January be on a Friday, the first Sunday will be on the 3d of January, therefore the dominical letter will be C; and the first Sunday the year after falling on the 1st of January, the Sunday letter will be A. In a common year all the Sundays in it have the same letter; but in leap-year, the additional day displaces the letters; therefore, if the first day in a common year fall on a Sunday, the next year it will happen on a Monday, and the next on Tuesday, and so on; and to prevent all the letters being displaced in a leap-year, the Sunday letter alone is altered.

Having mentioned the circumstance which occasioned one of the months to be named after Julius Cæsar; in justice to the abilities of Augustus, I cannot refrain from mentioning the circumstance which procured for him the like distinction, which was, his having ascertained the several elevations of the Sun above the horizon at different times of the year.—This he effected by means of the shadow of an obelisk 111 feet high, which he caused to be erected in the field of Mars for the purpose of this observation.

Ptolemy's astronomy, though founded on an erroneous system, served to give the observers of that age an idea of the apparent course of the heavenly bodies, as also to foretel natural events, and to bring geography to certain rules.

After the death of Ptolemy, speculative astronomy again began to decline, and at last was totally laid aside.

Historians inform us, that in the first ages of Christianity, the most learned Christians were wholly occupied in the important mission of instructing nations in the revealed religion, and in repelling innovators; which, added to the frequent changes of rulers, laws, and language, kept nations in a tumult unfavorable to science: that about the middle ages, the knowledge of our globe, history, and eloquence, were neglected; and that part alone of philosophy which belonged to logic and metaphysics, was in vogue: that negligent of the graces of elocution, they became rude in their manners and speech, and that their arguments were calculated rather to disgust and perplex than to convince. The latter of these assertions we may easily conceive must have been the consequence of the former, as, by experience, we know, that to confute without politeness and gentleness, is not the way to make our tenets respected or adopted.

It is said that these supercilious Arabian philosophers were shunned by all the world, and were considered as a public nuisance; as the doctrines they taught tended not to the service of either God or man, being subversive of all harmony and civilization.

Philosophy thus transformed, and stripped of all her fine embellishments, was rescued from total degradation in 1214, by some very few learned men, particularly by Roger Bacon our countryman, who about that time restored it to its native importance, clothing it with all that could render it lovely and respectable; so that it became an object of public esteem and suffrage.

In this century, the Emperor Frederic the Second caused Ptolemy's construction of the Universe to be translated from the Arabian into Latin.

In the year 1270, Alphonso, King of Castile, employed several learned men in the business of reforming Astronomy; and became himself an able astronomer. Charles, surnamed the Wise, gave great encouragement to this science. Copernicus, in the 15th century, re-established the ancient Pythagorean system, which admitted that the Earth might move round the Sun, by which the constitution of the heavens was again brought to natural and certain principles.

It was Gallileo who chiefly introduced telescopes into the use of Astronomy, in the year 1610, and by that means discovered the satellites of Jupiter, the phases of Saturn, the mountains of the Moon, the spots on the Sun, and the revolution of the latter on his axis; discoveries which opened a wide field of inquiry and speculation.

The immortal Newton was the first who demonstrated, from physical considerations, the laws that regulate all the motions of the heavenly bodies, as well as of our Earth; which set bounds to the planets orbits, and determine their greatest excursions from, and nearest approach to, the Sun, their grand vivifying principle.

He taught the cause of that constant and regular proportion observed by both primary and secondary planets, in their circulation round their central bodies, and their distances compared with their periods: he also introduced a new theory of the Moon, which accurately answers to all her irregularities, and accounts for them.

Doctor Halley favored us with the Astronomy of Comets, and, as I before mentioned, with a Catalogue of the Stars, together with astronomical tables.

Mr. Flamsteed, after observing the motions of all the Stars for upwards of forty years, gave some curious information on that subject, with a large catalogue of them.

Lastly, Doctor Herschel, whose opinion of the construction of the universe I shall give in the course of these Lectures, has very judiciously extended this field of science, and has discovered another planet belonging to our system : this gentleman's application to the science, and the liberal manner in which he has transmitted his observations, deserve great commendation.

I trust this short sketch of the origin and progress of Astronomy, and of the advantages it has procured for us, has not been unpleasing or useless, as the human mind must always feel satisfaction in tracing such things from their source to their utmost range ; and no doubt but the important inferences deducible from this epitome of ancient knowledge, must tend to enlarge the minds of those who have not been previously acquainted with these circumstances.

To preclude criticism, I must beg the historian to observe, that I did not think it necessary to my plan to introduce any thing of those times in which this science was not cultivated, or improved ; as to have related all the false systems that prevailed at different times, would have afforded but a mortifying retrospect, not tending to promote my grand design, in recording the speculations and works of past ages, which was to excite in my dear pupils a spirit of inquiry from the instances I produced of the advantages resulting from investigation ; which rule of selection has occasioned that want of connection necessary in writing the history of past ages, but not, I presume, in relating the history of the rise and advancement of astronomical knowledge, as it must necessarily have included matter foreign to the subject of these Lectures.

END OF THE SECOND LECTURE.

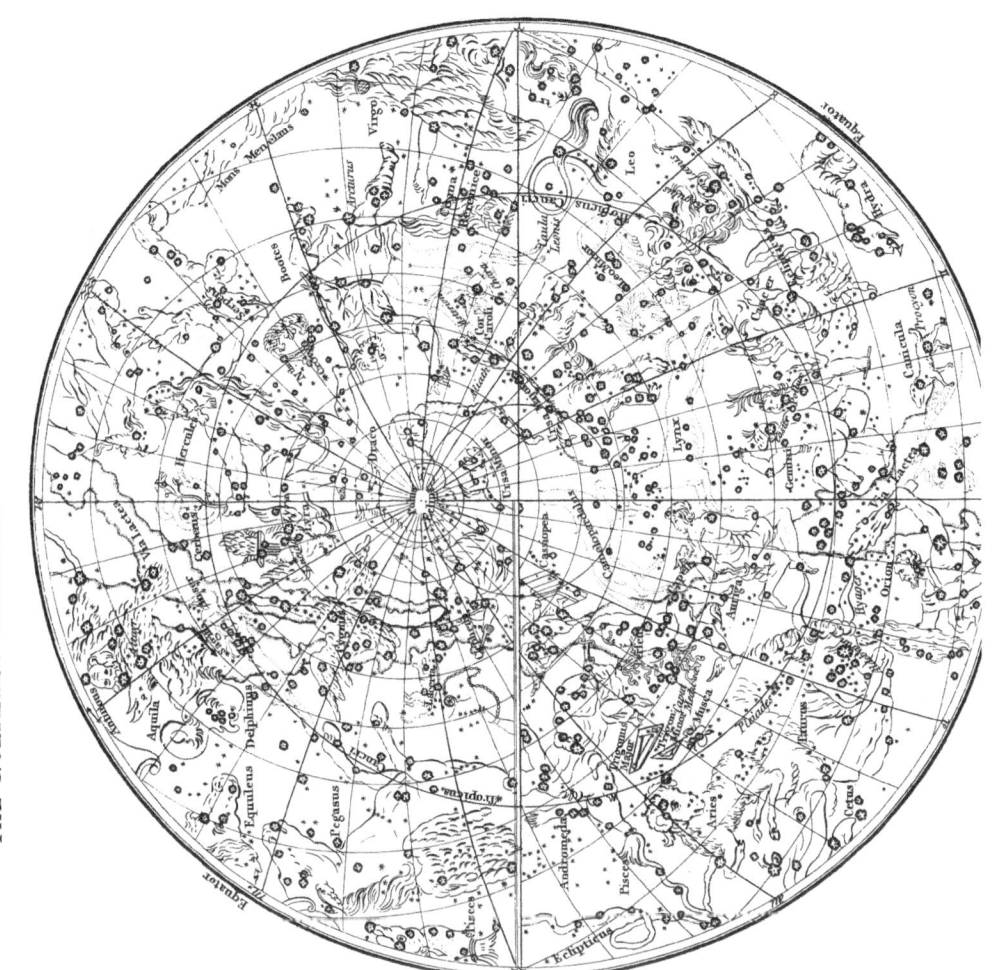

Plate IV.

THE NORTHERN CELESTIAL HEMISPHERE.

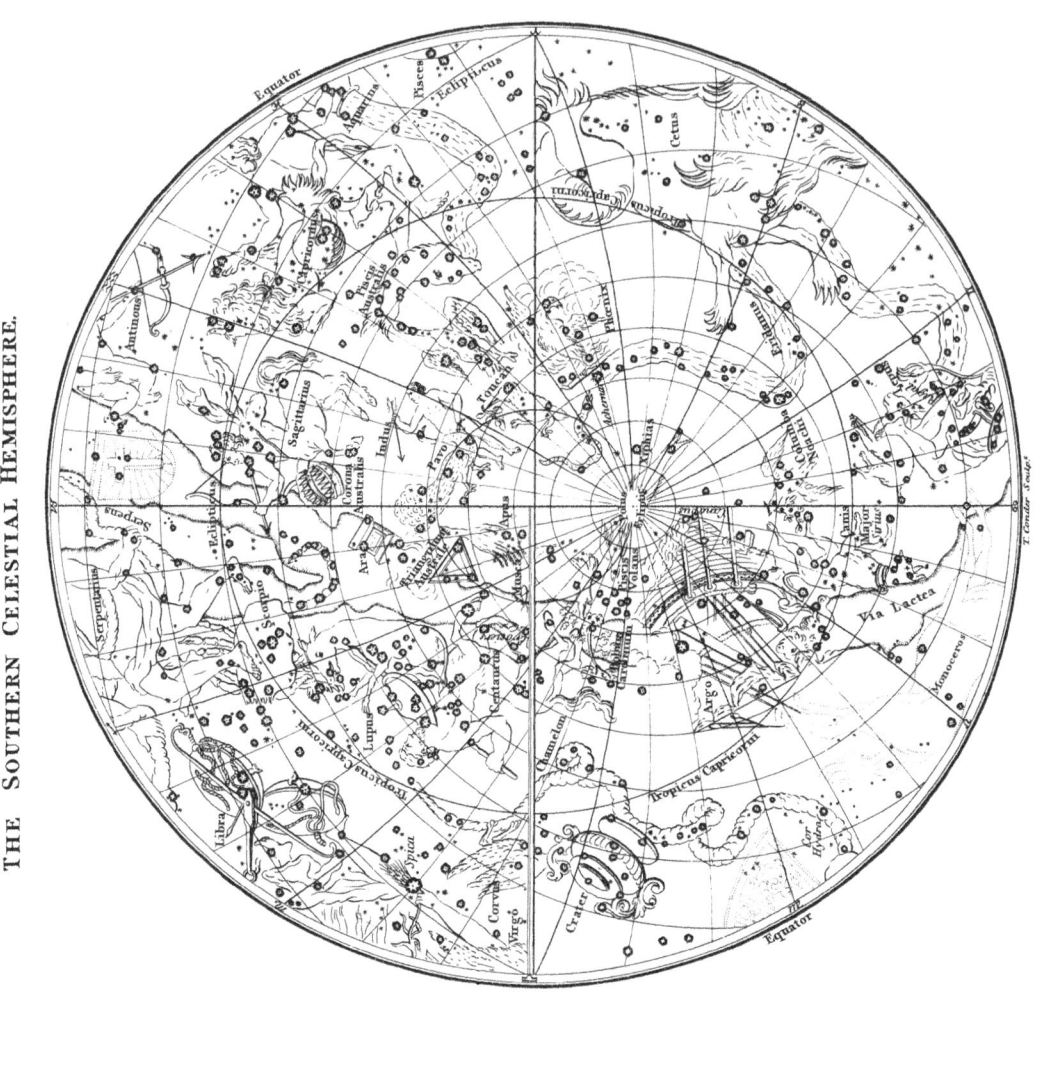

THE SOUTHERN CELESTIAL HEMISPHERE.

LECTURE III.

INTRODUCTORY ADDRESS.—OF THE FIGURE AND AFFECTIONS OF
THE GLOBE WE INHABIT, AND OF THE APPEARANCES OF THE
HEAVENS AND HEAVENLY BODIES IN CONSEQUENCE OF THE
LATTER.—THE CELESTIAL GLOBE AND ARMILARY SPHERE EX-
PLAINED, WITH THEIR USES IN REFERRING TO CELESTIAL
AND TERRESTRIAL PHŒNOMENA.—ZONES.—CLIMATES.—GE-
OGRAPHICAL DEFINITIONS, &C.—THE CAUSES OF THE VICIS-
SITUDE OF SEASONS AND CLIMATES.—A PHYSICAL DEFINITION
OF THE PRECESSION OF THE EQUINOXES; AND THE EFFECTS
OF THE SUN'S RAYS FALLING WITH GREATER OR LESS OBLI-
QUITY ON OUR EARTH.—MORAL REFLECTIONS AND INFER-
ENCES.

———————————— O, Spirit! that dost prefer
Before all temples th' upright heart and pure,
Instruct me, for thou know'st: thou from the first
Wast present; and with mighty wings outspread,
Dove-like, satst brooding on the vast abyss,
And mad'st it pregnant: what in me is dark,
Illumine; what is low, raise and support:
That to the height of this great argument
I may assert Eternal Providence,
And justify the ways of God to man.   MILTON.

Do not your breasts glow with the ardour of this divine poet,
when you reflect on the wonders which present themselves in heaven above, and on the earth beneath? they surely must.

That mind must be devoid of all sensibility that can behold the
resplendent glories of the heavens—feel the vivifying power of the
Sun—reflect on the advantages derived from transmitted and reflected light—examine into the diffusive excellence of the work

H

of creation—compare its general connections, inferring from thence the consolatory conviction of an all-wise, all-powerful, and all-merciful Creator, whose superintending and preventing goodness is never ceasing, is never diminished, and will endure for ever—without the warmest emotions of joy, gratitude, love, and admiration!

Cherish, my dear friends, this transport; it will elevate your minds above those low pleasures of sense, which debase human nature, and will animate you to an imitation of those graces and benevolent emanations, so conspicuous in the dispensations of Providence, which the contemplation of such sublime and interesting subjects must naturally tend to excite in your breasts.

Having in the two foregoing Lectures taken a review of such parts of the Astronomy of past times as was adapted to our purpose, I shall in the following ones endeavour to inform you of the principles of this noble science, and lead you on to consider the phœnomena which, in its present improved state, it so clearly elucidates.

The part of creation which claims our first attention, being immediately connected with us, is the globe we inhabit.

Our Earth is richly endowed with all things necessary to our temporal existence, both for the support of our animal nature, and also for the gratification of those senses we are furnished with. By its compound motion, we enjoy the grateful returns of day and night for the purposes of labour and rest, and the pleasing vicissitude of seasons. The former motion is rotary, and is called the diurnal motion, being that of the Earth round its axis, which is performed from west to east in twenty-four hours, occasioning the apparent motion of the Sun and the whole heavens, from east to west, in that time.

By the motion of the Earth on its axis, or rather on that imaginary line passing through the centre of the Earth round which it revolves, all the celestial bodies are alternately presented to our view, and hid from it, being disposed in respect to the appearances of things to us, in a concave sphere (though not actually so) which seems to encircle our Earth.

The other motion of our Earth is progressive, being performed round the Sun, and is called the annual, as the time it takes in performing its intire revolution round that luminary is called a year.

We have reason to suppose that our Sun, with its planets, forms only one link of the great chain of the universe, and that all those beautiful luminaries, the fixed stars, are equal in size and glory with our Sun, each serving to enlighten a system of planets in the same manner that our Sun does those of his system.

The distance of the fixed stars from our Earth must be very great, as there is no visible alteration in their positions, with respect to each other, when viewed from different parts of the Earth's orbit, that can be referred to their distance from us; so that the diameter of the Earth's orbit, though equal to 192 millions of miles, is but a point in comparison of their distance from us.

What an enlarged idea does this convey to us of the majesty of the great Creator !

Well might the enraptured poet exclaim,

> When all thy wonders, oh my God !
>   My rising soul surveys,
> Transported with the view, I'm lost
>   In wonder, love, and praise.   ADDISON.

I have informed you, that the figure of our Earth is globular, although not a perfect sphere; and have accounted for its deviation from that form.

The Earth being nearly spherical, and being impelled forwards, must move round a line within itself, as we observe all bodies of that form to do, when impelled forwards. The extremities of this imaginary axis are called the poles of the Earth; that pointing towards the part of the heavens denominated the north, is called the north pole, and that directed towards the opposite point, is called the south pole.

The change observed in the appearance of the visible starry hemisphere, I have before informed you, led to the discovery of the star now called the pole star, from its being situated near the point answering to one extremity of the axis of the diurnal motion; this point is considered as stationary in referring to the diurnal motion of our Earth; and when we transfer it to the heavens, it is equally so, on account of the immense distance of the polar star from us.

The elevation of this star above the horizon, differs according to the latitude of the place it is viewed from. The further we travel north, the higher will this star appear above the horizon; if we proceed southerly, it will become less elevated, till we shall at last lose sight of it, and another star will present itself in the south, near the other extremity of the axis of our Earth; but there is no star so near the line of motion in the south as in the north.

Although I am not fond of adverting to erroneous opinions, yet Ptolemy's system of the universe must not escape observation, as in refuting the errors of it, we establish the present known system of the world, and manifest the absurdity of his.

The falsity of the Ptolemiac system will be best understood, by comparing it with the known phœnomena of the heavens.

He supposed that our Earth was fixed in the centre of the universe; and that the Sun, with the other heavenly bodies, revolved round it once in twenty-four hours from east to west; also that they had different periodical times,—which latter he was obliged to admit, in order to account, as he imagined, for their annual changes, and various appearances.

This mistaken notion of things does not seem extraordinary in one who considered appearances only, without mathematical aid;

by which we have been brought to form just ideas of things so remote, and without which, united with the known laws of gravity
and attraction, we never could have attained to such sublime truths,
our senses alone being very inadequate to judge of the real motions
of the heavenly bodies, or of our Earth.

When a ship leaves the shore, if the sea be calm, the passengers
in the vessel do not perceive its progressive motion, but the objects
on land appear to them to move: thus, reasoning by analogy, we
can conceive, that our Earth moving through the subtile etherial
medium, where it meets with no sensible resistance; if it has a
motion, that motion cannot be perceived by us. Again, when we
consider that there is no such thing in nature as a larger body revolving round a less, as its centre of motion; we are convinced,
that as the Sun is known to be so much larger than our Earth, and
all the other planets of our system, that it must be the body included within the orbits of all the rest, having a less motion, or
describing a less circle round the centre of gravity of the whole
system, than any other body belonging to it.

It has been suggested by those who would not be convinced to
the contrary, either by reason or experience, that if our Earth had
a progressive motion, a stone dropped from any considerable elevation, would not fall directly perpendicular, as the body over
which it was dropped must have advanced considerably in the time
of its fall; not reflecting that a body when impelled by another
body in motion, partakes of the motion of that body.

The annual progressive motion of our Earth is equally deducible
from observation of the heavenly bodies;—for as its diurnal motion
is referred to the apparent motion of the Sun from east to west, so
is its annual revolution referable to that motion of this luminary,
discoverable from its rising and setting at different parts of our horizon, and traversing a certain space called the zodiac, in a year.

That circle which terminates our view, on whatever part of the

Earth we are placed, is called the horizon. This appears to divide the great expanse into two hemispheres; that above our horizon at the time, is called the visible hemisphere; and that hid from our view, the invisible.

Before the Sun rises above our horizon, a faint light proclaims his approach, which advantage we derive from our atmosphere, as I have previously informed you, by its reflecting the rays of light to our eyes, long before we feel their direct influence.

Were it not for this property of the atmosphere, we should not only lose the advantage of the mild approaches of this luminary, by which our sight would be impaired, but likewise when he was visible to us, all objects would be dark, excepting those which received the direct impulse of his rays.

The density of the atmosphere decreases with its height, as has been proved by experiments. A barometer being placed at the foot of a mountain, has stood, suppose at 29 inches, by the great pressure of the air, which must have been occasioned by its density and pressure; but taken to the top of the mountain, it has sunk to 25 inches, and that by a progressive gradation in its ascent. On comparing the diminution with the distance in the foregoing observation, it has been computed that the height of the atmosphere is about 20 leagues above our Earth; but as the atmosphere seems to be indefinitely extensible, I think that its real height above our Earth cannot be accurately ascertained.

This fluid, which surrounds our Earth and revolves with it about the Sun, is what we call sky; which, although of considerable density, yet it transmits the light of the Sun to us, and allows us to view the glories of the heavens in all their splendor and harmony, unless obscured by the gross vapours of our Earth.

The effects of the accumulation I am speaking of, was a great impediment to navigation, whilst the polar star was the only guide to mariners; as when it was thus obscured, they had no other by which they could steer their wandering vessels. — At length it pleased the Almighty, by one of those circumstances miscalled casualties, to afford them a more constant guide, by imparting to mankind the knowledge of that property of the magnet by which it points to the pole of the world.

The essential properties which cause this phœnomenon have not been ascertained. It is thought to be produced by a subtile fluid, which fluid is considered as being composed of two kinds of elements, united by affinity; and that these elements have a greater tendency to each other than those of the same kind. Excepting the aforementioned tendency, their attractions follow the laws of gravity, and this fluid has affinity with the particles of iron.

After having engaged the attention of the most celebrated Philosophers, the essential properties of this subtile fluid, like those of light, gravity, and electricity, remain undefined; from which we may conclude, that we are let to know no more than is necessary for our comfort and convenience in this life, and that there is a boundary we shall not be permitted to pass, till we are translated into that pure state in which we shall be able to behold the works of Providence as they really are — and be made to understand those holy mysteries, which, for wise purposes, are now hid from us; for, as the excellent Dr. Blair remarked, in a fermon I once had the pleasure of reading to you, " This obscurity is our happiness; as were we permitted to see the beauties of the celestial regions, and to comprehend the excellence of divine things, we should disrelish our present enjoyments." Let us then be satisfied with the usefulness of the effects, and in tracing the benevolence of the works of God, as we have good reason to be — although we cannot develope the glories of the heavens, or understand all the energies of the Divine Mind. — But to return to the magnet, and its use in navigation.

The north pole of this curious and important phœnomenon points nearly to that spot which is supposed to represent the northern extremity of the axis of the Earth's motion; yet it very seldom points exactly to it; so that the magnetic meridian seldom answers exactly to the meridian of any place on our globe.

The meridian of a place is that circle which passes over it, and through the poles of the world. The Sun is on this line at twelve o'clock at noon each day, which with Astronomers is called the beginning of the first hour of the day, but by others the middle hour of the twenty-four. The variation of the needle from this line is different at different places on our earth, both by fea and land; it is also continually varying at the same place; so that although in the year 1580 it was 11 degrees east at London, it is now 23 degrees and a half west at that city. The variation is always reckoned from the north pole, east or west.

The directive power of the magnet was applied to the purposes of navigation in the fourteenth century, yet its variation was not discovered till the latter end of the fifteenth century; prior to which it was supposed always to point due north and fouth.

Columbus, who discovered the deviation, did not perceive the extent of it, imagining it to be invariably the same at the same places; and it was not discovered till the year 1625 to be different, at different times, at the same place.

Various are the theories advanced to account for these irregularities; but all being mere conjecture, they are by no means satisfactory. However, I shall state some of the most popular ideas concerning them, although it is impossible for me to ascertain their authenticity.

Supposing our Earth to be encompassed by a magnetic fluid; magnetic needles placed on its surface, freely suspended, would have different directions at different places; which is conformable to experience—and the apparent irregularities in the variation of the needle, must be occasioned by the situation of the magnetic poles.

If the magnetic poles agreed with the poles of the world, there would be no variation; so that the magnetic needle would point due north and south continually.

If the axis of the magnetic poles passed through the centre of the Earth, it would be easy to assign the quantity of variation at each place: but this is not the case; therefore, to account regularly for the variation, it is necessary to know the exact situation of the magnetic poles, their number and force, also their distance from the real poles; whether they shift their places; and, if they do, the quantity of motion every year—circumstances which philosophers have in vain endeavoured to discover—an uncertainty which is some impediment to navigation.

The dip of the needle, or its depression below the horizon, is also subject to a variation, and is said to be not less than 72 degrees now at London.

The ancient Astronomers divided the horizon into four parts, and distinguished only four points by names, which they called the cardinal points, as being the principal; these were placed at the intersections of the meridian and prime vertical circles with the horizon. Those of the meridian with the horizon they called north and south; and the other two, at right angles to the former, the east and west.—These four are still called the cardinal points, and the intermediate the collateral.

The compass, or the boundary of our view on all sides horizontally, is now divided into thirty-two parts, or points, and their distances from their primaries are expressed by degrees.

The point immediately over our heads is called our zenith, and refers to the highest point of the apparent celestial sphere; it is equi-distant from every part of the horizon, and is reckoned 90 degrees, or a quarter of a circle, every way from it.

I

The nadir is the lowest point of the celestial sphere, and is hid from us, being that point directly opposed to our zenith.

Every circle, large or small, is by mathematicians divided into 360 parts, which they call degrees; 90 degrees, or a quarter of a circle, is usually employed to measure all angles formed between a line perpendicular to the horizon which goes up to our zenith, and all other lines that are not below our horizon.

Our Earth being a spherical body, the limit of our view must change at every change of our situation on this globe. A spectator on the Earth at A, *fig. 4, plate* 1, has his view bounded by the line from B to C; if he travels to D, his visible horizon will be at E F.

That circle which we suppose situated directly between the north and south points of the compass, and passing through the east and west, is called the equator of our Earth; because it equates or divides the globe into two equal parts: that on the north side of this circle is called the northern hemisphere, and that on the south side of it is called the southern hemisphere.

The poles of the world are the poles of the equator, and the zenith and nadir may be called the poles of the horizon; as the poles of every circle of a sphere are those two points which are at the greatest distance from the plane of that circle; and the poles are always 90 degrees from the circle every way.

The highest point to which the Sun ascends, at which he arrives nearly at twelve o'clock each day, is at the meridian; if from this spot we describe a circle through the north and south points of the compass, and passing through our zenith, we shall call that the meridian of the place of observation.—Every place on our Earth has its particular meridian passing over its zenith, and through the north and south points of the compass; or from the north and south poles of the world.

We do not change our meridian by travelling north or south, but only our zenith; if we travel east or west, our meridian is changed by every removal; all the meridians intersect each other at the poles, as shewn *fig.* 4, *plate* 1.—The equator being every where equi-distant from the poles of the world, it divides every meridian into two equal parts.

The poles of the heavens are formed by a continuation of the axis of our Earth; and, in like manner, the circles I have been describing are transferred to the celestial concave; so that when I am describing celestial phœnomena, and speak of the equatorial, horizontal, and meridional circles, I transfer those of our Earth to the Heavens; which, although imaginary, serve as fixed points for observation, as it is to them, together with the ecliptic, that we refer all the motions, &c. of the heavenly bodies.

As the horizon of a place distinguishes night and day at that place, the length of the day and night must differ at every place; for as each has a different zenith point over it, the horizon must also be different; as is evident from an inspection of *fig.* 4. *plate* 1.

The apparent rising and setting of all the heavenly bodies are referred to the horizon; and the higher they are elevated above it, the longer they stay with us.

Mathematicians denominate all those circles great, the diameter of which passes through the centre of the sphere, and the planes of which divide the globe into two equal parts.

The equator, the meridian and horizon are great circles of our globe, and together with the ecliptic, the great circles of the celestial sphere. Small circles divide the globe into two unequal parts; such are the parellels to the equator of a terreſtial globe, and the parallels to the eliptic of a celestial sphere.

In referring our horizon, or the boundary of our view, to the heavens, we regard only our sensible horizon; our rational horizon, or that, which if our globe were of sufficient magnitude in comparison of the great expanse, would as it were cut our globe in halves, would differ from our sensible one.

Let A B C D, *fig.* 5, *plate* 1, represent our Earth, and E F G the heavens, which is near enough to the body A B C D to form an angle with it, measured by *m n*, when we compare the line drawn from its centre, and the one from its surface, as from A to *m*; but if we compare the lines drawn through the middle of the Earth to the supposed heavens at O, in their more distant situation, as represented by the semi-circle H I K, with the other line drawn from the top of the Earth to O, we shall perceive that these lines, on account of the great distance, terminate the same, meeting in a point at O;—this is the case in fact, our Earth being but a point in comparison of the distance of the fixed stars; as is evident by our always seeing the polar star in the same point of the heavens, notwithstanding the annual motion of our Earth.

I will now endeavour to explain some of the phœnomena of the heavens by means of the celestial globe, represented *fig.* 1, *plate* 2. It is called the celestial sphere, because on it are delineated those stars we see, and those circles we imagine, in the concave sphere of the heavens. But being on a convex surface, they appear in a reverse order, and therefore, in transferring them to the heavens, we must suppose ourselves situated within the centre of it; as is represented by the armilary sphere, *fig.* 1, *plate* 3, in which our Earth is in its proper situation in respect to those circles of the heavens represented by the armilary part of it. I shall refer you to each occasionally; when speaking of the imaginary circles of the celestial expanse, the armilary sphere will exhibit these phœnomena most clearly—when referring to other celestial circumstances, the celestial globe will best answer our purpose. And by their several uses I hope to make you understand all the apparent motions and changes of the heavens, so far as relate to the circles of the sphere, and to the real annual and diurnal motion of our earth. Preparatory to which, I will explain the rationale of the celestial globe.

This globe, being duly balanced, turns with facility round a pin, which serves it for an axis, as our Earth does on its axis, or line of gravity; which motion exhibits the apparent motion of the heavens, occasioned by the diurnal motion of our Earth.

The broad circle CD of a celestial globe always is used to represent the horizon of any place on our Earth, when the latitude of the place is brought to cut the edge of the horizon.—This I shall explain more fully to you hereafter.

The horizon of this globe has four separate circular spaces delineated on its surface, the innermost of which is divided into 360 degrees, and four quadrants; the numbers on the latter begin at the east and west points of the compass, being counted each way to 90 degrees, which is at the north and south points.

The next circular space contains the 32 points of the mariners compass; the next, the 12 signs of the zodiac, which are divided into 30 degrees each; and these are characterized by the names, signs, and figures, belonging to them. The outer circle of the horizon contains the months and days; the former are divided from each other, and the latter numbered.

The brass graduated circle which crosses the globe, north and south, is called the general meridian, because it serves to shew the meridian and zenith of every place on our Earth.

This circle is divided into quadrants, each of which you know must contain 90 degrees. The numbers on two of the quadrants begin at the circle on the globe which represents the equator, and increase towards the poles; being used to shew the distance of any star or circle of this globe from that imaginary circle of the heavens which it represents. The other two quadrants have their numbers counted from the poles towards the equator, being used for the purpose of fixing the globe to the latitude of any place, which is always estimated by the elevation of the pole above the horizon of the place.

The Sun's geocentric place for 12 o'clock every day in the year, is known by reference to the days of the month, and signs of the zodiac, on the horizon of this globe; but as his splendor at that time prevents our seeing those signs in which he is, we are obliged to defer the observation till that luminary has withdrawn his beams, when we shall perceive the sign directly opposite to that in which he was at 12 o'clock at noon of our day, at 12 o'clock at night on our meridian.

The graduated side of the brass meridian of a celestial globe should face the east, as the globe should always be turned from east to west, to represent the apparent diurnal motion of the Sun and heavens.

In Adams's globe, which is the one represented, *fig.* 1, *plate 2,* the equator is used for the hour circle, which is, I think, very proper; as the time that the apparent revolution of the heavens is performed in, is 24 hours, which you know is occasioned by the real motion of our earth on its axis in that time, in a contrary direction, or from west to east.

Over this circle on this globe, which serves for an hour circle, is placed a semi-circular wire that carries two indices, which serve to shew how long any planet is in performing any part of its revolution, and when it will be fit for observation; also the same of the apparent motion of the stars, and the time of their being above our horizon, as I shall explain in the problems. The hours are placed below the degrees of the equator of this globe.

The equator of our Earth is called the *Line* by mariners; and when they pass over it, they say they have crossed the line.

The Sun moves in this imaginary circle of the heavens on two days only in the year; the one in spring, and the other in autumn, on which days the day and night are equal all over the world;

and hence these intersections of the ecliptic, or Sun's path, with the equatorial, are called the equinoctial points.

The ecliptic, or apparent path of the Sun in the heavens, forms with the equatorial (or imaginary circle, which is supposed to be exactly between the north and south pole of the world) an angle of 23 degrees and a half.

The ecliptic on the globe is divided into twelve equal parts, each of which contains 30 degrees, corresponding to the space occupied by each sign of the zodiac in the heavens. The Sun always appears among the stars described on this zone, and to advance among them easterly, a space nearly equal to a degree each day, going through all the signs exactly in a year.

The equinoctial points, or intersections of the ecliptic and equator, are at the beginning of Aries and Libra. The first of Cancer and the first of Capricorn, are called the solstitial points, because when the Sun arrives at either of them, he appears to be stationary for several days, or to arrive nearly at the same elevation in the heavens above the horizon. When he is in one solstice, which is at that circle parallel to the equator called the tropic of Cancer, he makes the largest angle with the horizon in northern latitudes, which at that of London, being 51 degrees and a half north, makes the longest day happen on the 21st of June. When he is at the other solstice, which is at the tropic of Capricorn, he makes the least angle with the horizon of London, having the least elevation above it, and the shortest day is on the 21st of December at that city.

In southern latitudes it happens in a reverse order, their longest day being when the Sun is at the tropic of Capricorn, and their shortest when he is at the tropic of Cancer. In both cases, the length of the day must increase with the degree of latitude, as the higher the Sun rises above the horizon of a place, the longer will he remain above it, and therefore the length of the longest day, and of the shortest, at every place, will differ from those of every other place, according to the latitude of each.

The 24 circles on the celestial globe, which intersect each other at the poles of the ecliptic, are used to shew the latitude of the fixed stars, which is reckoned by their distances from the ecliptic.

The latitude of the stars is counted from the ecliptic, the most important circle of the heavens, being that in which the Sun always appears in glorious majesty;—and in reference to his apparent motion we speak of the poles of the ecliptic.

The Sun has no latitude, being always in the ecliptic; and to speak of his longitude is not so proper as his place in the ecliptic, because his motion is only apparent.

The eight circles drawn parallel to the ecliptic on this globe, are to shew the latitude of the planets, those never deviating from the ecliptic more than the space included within the exterior of these eight circles on each side of it.

The longitudes of the planets are reckoned on the ecliptic, and eight parallel lines on each side of it, beginning at the first degree of Aries, and counting eastward.

The longitude of the stars is also counted on the ecliptic, and all the parallels to it.

The difference between the longitude and latitude of places on our Earth, on a terrestrial globe, and those of the heavenly bodies on a celestial sphere, are these: The longitudes of the heavenly bodies are counted on the ecliptic and parallels to it, beginning from that part of it distinguished by the sign Aries; but the longitude of places on our Earth are counted from the equator and parallels to it; and the first degree, or where we begin to count from, in England, is at London, or at Greenwich; so that the celestial and terrestrial longitude differ from each other, the former being settled at the first degree of Aries, and the latter at London, which is not in the same spot. The latitude of the heavenly bodies also differs from that of terrestrial places; the first being counted from the ecliptic, but the second from the equator.

The distance of any heavenly body from the equator, measured upon the meridian, is called its declination, which answers to the terrestrial latitude.

The brass semi-circle fixed to the poles of the globe, (as represented by G I, *fig.* 1, *plate* 2) renders the working of problems of declination more convenient than by the general meridian, it having a sliding piece of brass (H) on it, which may be fixed to any spot on the globe.

The Sun's declination, as seen on the celestial sphere, is the same as the latitude of the place to which he is then vertical, as seen by a terrestrial globe; from which you perceive, that parallels of declination on the celestial sphere are the same as parallels of latitude on the terrestrial globe, both being those parallel to the equator.

Those two parallels to the equator called the tropics, (E D and C F, *fig.* 2, *plate* 2) are 23 degrees and a half distant from it on each side, being situated at the Sun's greatest deviations from it, north and south.

As the Sun is advancing towards Cancer from Aries, he appears more and more northerly every day above our horizon, when he arrives at the meridian; so that he is more and more elevated above it each day till he is arrived at Cancer, which is his most northern situation. After arriving at his highest situation, he descends from it in the same manner, appearing lower or nearer to the horizon each day at noon, till he arrives at Capricorn, his most southern situation.

This apparent motion of the Sun backwards and forwards, between north and south, has occasioned the parallels at which he arrives to be called tropics, with the additional name of the sign of the zodiac which touches those parallels in which the Sun appears

K

at the time, one being called the tropic of Cancer, and the other that of Capricorn.

The surface of our Earth is divided into five zones, which are denominated according to the degrees of heat or cold felt at those parts, *fig. 3, plate* 2.

The space between the tropics and polar circles are called temperate zones; the breadth of each is 43 degrees 4 minutes.

The torrid zone is all the space between the tropics, the breadth of which is 46 deg. 56 min. The equator divides it into two equal parts, containing 23 deg. 28 min. each. By the ancients this zone was thought not habitable, on account of the great power of the Sun on those parts. They did not reflect on that part of the constitution of animal nature, which is provided by the Deity, to moderate the heat of all climates, or to accommodate itself to them.

The spaces between the poles and the polar circles are called frigid zones; the breadth of each segment between them is 46 deg. 56 min. as from A to B, or 23 deg. 28 min. every way from the pole.

The inhabitants of some part of the torrid zone have the Sun pass through their zenith twice a year, at which time the heat is excessive, by reason of the great quantity of the Sun's rays they receive at that time. The people inhabiting the parallels to the equator, as C D, *fig. 3, plate* 2, within the torrid zone, are called Amphiscians, from Amphiscii, which signifies about and shadow, because these people have their shadow at noon on opposite sides of them at different times of the year.

Those who live immediately on that part of the torrid zone, described by the equator, or on the centre of it, are called Ascii or Ascians, because at certain times of the year they have no shadow. In the torrid zone, the Sun rises and sets every natural day, because the distance of the Sun from the pole always exceeds the height of the pole in those latitudes.

In the temperate zones also, the Sun rises and sets every natural day, because the height of the pole is less than the distance of the Sun from the pole.

In the temperate zones, the least height of the pole exceeds the greatest distance of the Sun from the equatorial, and therefore the Sun never passes through the zenith of the inhabitants of these zones. Where the temperate zones join the frigid, the height of the pole being equal to the Sun's distance from it, when the Sun is in either of these tropics, the inhabitants of those circles see the Sun perform an intire revolution above their horizon, so that they have perpetual day; and when he is arrived at either, the other has perpetual night.

The people inhabiting the temperate zones, are said to be Heteroscii, which signifies that during the whole year they have their shadows at noon projected different ways from each other. We, who inhabit the northern temperate zone, as at F, *fig.* 3, *plate* 2, are Heteroscii to those who inhabit the southern temperate zone, as at G, and they are said to be the same in respect to us.

In the frigid zone, the least height of the pole exceeds the greatest distance of the Sun from the equator, the same as happens in the temperate zone, and therefore the Sun never passes through the zenith of the inhabitants of these zones. Every where in a frigid zone, the height of the pole is greater than the least distance of the Sun from the pole, so that in some parts of the Earth's annual revolution, it is at a less distance from the pole than the poles height above the horizon, at which times it does not set below the horizon, or even touch it, in a frigid zone.

At such parts of the Earth's orbit, where the Sun's distance from the pole exceeds the height of it, or the latitude of the place, the Sun rises and sets every day.

The people who live at the poles are called Periscians, from Periscii, because their shadows go quite round them in the same day, when the Sun does not set to them.

Astronomers and Geographers divide the Earth into 24 climates of half hours, and six of months.—The surface of our Earth between the equator and the polar circles, they divide it into 24 parts on each side of the equator, which are distinguished on a terrestrial globe by circles drawn parallel to the equator ; and may be seen, *fig. 2, plate* 3.—There is the difference of half an hour in the time when the Sun is above the horizon on the longest day at one parallel, and the next to it; therefore the days are longer by half an hour on one parallel than on the former, throughout the whole 24 parallels from the equator to the polar circles.

As the longest natural day at the equator is 12 hours long, at the polar circles it must be 24. In the latitude of London, which is on the ninth parallel from the equator, as at C D, *fig.* 2, *plate* 3, the longest day contains 16 hours and a half.

Between the polar circles and the poles, six parallels are drawn : on the first parallel, the day is one month long, and it increases one month at each, so that at the poles the day is six months long.

We may consider our Earth, in respect to its different inhabitants, as we do a ball of magnet, which by its power of attraction, when rolled in steel filings, supports them in all possible directions, even when turned with the greatest celerity—as it is thus that the Earth by its attraction draws every body on its surface towards its centre, and by its great attractive power retains them on its surface, notwithstanding the velocity of its motion on its axis.

Those people are called Antipodes, who are situated in direct opposition to each other on the Earth. Their hours, seasons, and days, are also in a direct opposite order, it being day with one, when night with the other. They are at the same distance from the equator, but on opposite meridians, as at A and B, C and D, E and F, *fig.* 3, *plate* 3.

The Antiecians are those who, inhabiting the same parallel and the same portion of the meridian, (only one north and the other south) have their hours the same, but their seasons different, as G and H, *fig. 3, plate 3.*

Those who are situated on the same parallel and same meridian, only differing by half circles of the meridian, as at G and I, or H and K, have their seasons the same; and it is midnight with one, whilst noon with the other.

There are three situations of the celestial sphere in regard to our Earth, which are easily shewn by either the globe or armillary sphere, and may be seen by *fig.* 4, 5, 6, *plate 3.*

All those who live under the equator are situated in a right sphere, because the poles are in their horizon, and their zenith point is in their equator. In which situation, the heavenly bodies rise and set right or perpendicularly to the horizon, and the Sun's apparent path is cut into two equal parts by it; which makes the days and nights equal in length to the inhabitants of the equator all the year through; the Sun being twelve hours above, and twelve below the horizon, during each revolution of the Earth on its axis. See *fig.* 4, *plate 3.*

A parallel situation, or sphere, is the situation of those people who inhabit the poles.—These have the equator in their horizon, one of the poles being in their zenith and the other in their nadir. In this position the stars never rise or set to their horizon, because the apparent motion of those luminaries is parallel to it;—so that they have but one day and one night all the year, each six months long. Those of the north pole have perpetual day for the six months that the Sun is on the northern side of the equator, and those of the south pole for the six months that the Sun is on the southern side of the equator; fee *fig.* 5, *plate 3.*

In our island we are situated in an oblique sphere, and so are all those who do not live under the poles or the equator; as to us and them, all the circles parallel to the equator make oblique angles with the horizon; so that there are as many oblique positions of the globe as there are positions from the equator to the poles. See *fig.* 6, *plate* 3.

In an oblique situation of the globe, the orbits of all the heavenly bodies make oblique angles with the horizon; which make the days and nights unequal, excepting at the equinoctial points, or intersections of the equatorial and ecliptic. At all other times of the year the daily circles of the Sun are unequally divided by the horizon of those who live in an oblique sphere; but when the Sun is in the equator, that circle being divided into two equal parts by the horizon, it is equal day and night—which happens about the 21st of March and September. All the rest of the year the days are longer or shorter than the nights, to the inhabitants of an oblique sphere. When the Sun is on the northern side of the equator, those situated in the northern part of our globe have their days longer their nights, while those in the southern have theirs shorter. When the Sun is on the southern side of the equator, those inhabiting that part have their days longer than their nights; whilst those on the northern have theirs shorter.

The polar circles have been called the arctic and antartic, and have those terms always affixed to them on globes and planispheres, though not with strict propriety unlimitedly; as by an arctic circle is meant the largest parallel, that is always above the horizon of any place in the northern hemisphere, including all those stars which never set to the horizon, but make a complete circle above it, and which is also called the circle of perpetual apparition, being never hid from our view.

By the antarctic circle is meant the largest parallel that is entirely hid below the horizon of any place, and which is the circle of perpetual occultation, never rising above the horizon of the inhabitants of the northern hemisphere.

All arctic circles touch the horizon in the northern point of the compass; and all antarctics touch the horizon in the southern point of the compass.

The polar circles do not answer for the arctic and antarctic to any of the inhabitants of our Earth, but to those who live under either of the tropics.—This is easily seen by fixing either of the tropics on a sphere to the zenith point, and then placing it in any other situation; as then you will perceive that the polar circles are the arctic and antarctic to the polar regions only, every other place having its arctic and antarctic circle in different parallels.

The cause of the groups of stars on the celestial sphere being included within familiar figures, you are already acquainted with; of which constellations you know there are twelve on that path or circle which the Sun appears to pass through, and to every part of which he is actually transferred by our observations, on account of the annual motion of our Earth. All the constellations on the north side of the zodiac are called northern, and those on the south side of it are called southern.

I have annexed to the plates of the celestial hemisphere a catalogue of the constellations, and remarked the principal stars in them by their names at full length.

The constellations are said to increase their longitude continually from west to east; contrary to the apparent motion of the heavens, by which the first star in the constellation Aries, which 1900 years ago was at the vernal equinoctial point, is now removed about thirty degrees from it, which is the whole constellation, or one-twelfth part of the zodiac; to conform to which circumstance delineators of celestial spheres have placed the character or sign only of the constellation at that point, having removed the star that used to occupy that station thirty degrees from it; so that

if we want to find the place of a particular star in any of the constellations of the zodiac, we look for it within the figure of it;
but if we wish to compare the circles of the sphere with any circumstance relative to the planets, we look for the character only,
which is called the sign, on the ecliptic circle, and not at the figure
or constellation itself.

This change of the longitude of the stars, or the retrograde motion of the intersection of the ecliptic with the equator, is called
the precession of the equinoxes, which motion is found to be about
50 seconds in a year, and is said to be owing to the attraction of
the Sun and Moon on the protuberant matter about the equator.

The opinion of Sir Isaac Newton touching the physical cause of
the precession of the equinoxes, is, that it arises from the broad
spheriodical figure of the Earth; which is a natural inference, as
the accumulated matter about those parts must occasion a greater
degree of attraction both by the Sun and Moon.

The variation which has been discovered by Dr. Bradley in the
quantity of the change of the equinoctial points, one year with
another, is ascribed to the variation in the attraction of the Moon,
as the plane of her orbit is sometimes above 10 degrees more inclined to the plane of the equator, than at another; therefore that
part of the whole annual motion which depends on the action of
the Moon, must vary in its quantity, although that which depends
on the attraction of the Sun continues the same, that luminary
keeping nearly the same inclination to the equator every year.—
Therefore, although the mean annual precession proceeding from
the united influence of the Sun and Moon, is 50 seconds, yet the
apparent annual precession sometimes exceeds that mean quantity,
according to the various situation of the Moon's nodes.

The idea of the obliquity of the ecliptic to the equator being
variable, seems to have arisen from the change observed in the latitude of the fixed stars, as well as from the actual diminution of

the angles they form between them, discoverable by observation with good instruments, and is also admissible on the principles of attraction referred to the variable inclination of the Moon's orbit, which must affect its action on the redundant matter about the equator, both as to its longitude and latitude.

The precession of the equinoxes is performed in a contrary order from the apparent annual motion of the fixed stars, being from east to west; so that the places of those luminaries, although immoveable, are transferred to a more western situation.

As great changes have happened in the situation of things on our Earth, some have imagined that the position of it, in respect to the fixed stars, may also have been variable; but this latter idea I do not think admissible, there being no good grounds for such a conjecture; because in old records, ancient historians, as Herodotus, in relating the events which happened in the time of the Egyptian Astronomy, have made some inferences totally inconsistent with our better knowledge of the physical principles of that science.

It was impossible for Herodotus, or any of his contemporaries, to form accurate ideas of circumstances preceding their time, having no better authority for the events they related, than that of the hieroglyphical traditions of the Egyptians; and how liable those were to be misconceived, I have already informed you. It requires strong judgment and deep reasoning to digest the writings of Herodotus, as in the events he relates, fiction and truth are so blended, that human comprehension is scarcely equal to the task of separating them; and for this his veracity is not to be impeached, as it originated in the materials from which his work was composed.—The fault was in their ambiguity, not in his skill in compiling and uniting; in both of which he evinces great discernment and deep erudition He has great merit as an historian, by not attempting to mislead; as, on the contrary, he frequently intimates a doubt of the proba-

I

bility of those circumstances, which, as a faithful historian, he is obliged to relate.

The Sun's apparent motion is unequal; i. e. the motion of our Earth in its orbit is sometimes swifter than at others, on account of the elliptic form of its orbit, which causes it to be nearer to the Sun in one part of it than in others; the former happens in winter, so that there are eight days more from the vernal to the autumnal equinox, than from the autumnal to the vernal. The summer's heat and the winter's cold, are occasioned by the different obliquity of the Sun's rays falling upon our Earth, and the different duration of that luminary above the horizon. In the former season, they fall with less obliquity, and therefore more of his rays are received on particular parts of our globe: in the latter season, they fall with greater obliquity, therefore less of his influence is received, and he continues a less time above the horizon also.

Fig. 7, *plate* 3, will convey a clear idea of what I have been asserting. Suppose A B C D to represent our Earth, and H I the horizon of a spectator at the latitude of London ; K to be the elevated place of the Sun above the horizon in summer, and L that above it in winter. You observe that the rays of the Sun in the former case fall with less obliquity, and therefore the same quantity fall in the small space of our globe, *m n*, as do from the Sun in the latter case on the large space of it, *o p*, on account of their greater obliquity in the latter case. You also observe that they pass through a greater quantity of dense atmosphere in the winter. by which also we may suppose their power is somewhat diminished or weakened by the reflective and refractive power of that medium.

We observe that the Sun rises at different points of the compass, or at different parts of the horizon of any place we live at, at different seasons of the year; this occasions him to make sometimes a larger angle with it than at others ; of course he must be longer above our horizon, making a larger circuit above it ; and his rays are less oblique than when he makes a less angle with it.

From December till June he is continually increasing his diurnal range northerly, and of course rises higher each day than the preceding one: thus the days increase in length during that time.

From June till December he appears to return back towards the south, making continually less and less stay above the horizon, and of course the days then decrease.

You now perceive the cause of the pleasing vicissitude of seasons;—that when the Sun rises high above our horizon, by the continuance of his influence, and the less obliquity of his rays, the fruits of the Earth are ripened; and that after performing this salutary office, he retires from us by degrees, and bestows the like benign influence on other countries.—At our time of need he returns again, and thus is continually and abundantly aiding every part of the habitable world.

How grand the design, and how perfect the accomplishment of as much as we have already explored! Then what must be your admiration, astonishment, and gratitude, as we advance in the scrutiny of the wise administration of Providence?—It will exceed all that we can form an idea of, and contain more than we ever can comprehend; for the ways of the Almighty are unsearchable, although his attributes are as clear as noon-day.

Religion being the firmest support of happiness, I wish to fix your attention on subjects that will lead you to the attainment of it; and shall therefore occasionally display the attributes of the Deity, deducible from the subjects of these Lectures, by which I shall fortify your minds against the shock of accident, and the terror of calamity;—raising your hopes on that solid basis, which time nor circumstance can never shake.

Irreligion amongst the unreflecting is so prevalent, that both my duty and affection prompt me to enlighten your minds, that you may not be blinded by passion or ignorance; but by having your understandings cleared from those mists, which hide so many things from common observation, you may be established in faith and goodness.

I regret that these subjects* are not more generally understood; but as you, my dear young friends, will sensibly feel the great advantage resulting from the instruction they convey, you will, I hope, endeavour to use them for the benefit of others, but without vanity or ostentation—rather recommending these studies for the delight they afford, than as being indispensibly necessary;—unless you perceive any of your fellow creatures doubtful of an over-ruling Providence, or deviating from the paths of moral rectitude, then indeed the conviction such subjects as these bring with them, renders them indispensible; and it becomes your duty (if you have the opportunity) thus to convince them of their error, and to re-claim them; the arguments these studies furnish you with, being more likely to effect that desirable purpose, than all the remon-strances you could advance, being truths which cannot be confuted by the most subtle arguer; for, like gold in the furnace, the more they are tried, the brighter and more beautiful their intrinsic worth will appear. Yet disgrace not, as some do, the cause you mean to serve, by a pharisaical conduct, by a display of your own me-rits or deserts; such is the infirmity of human nature, that none are exempt from error; all we can do is to strive against it; and although we may, by the advantages afforded us, avoid the more flagrant faults, or perhaps misconceptions of our neighbours, yet we can deserve no praise,—although we should the severest blame, —if we shut our hearts against conviction, and acted contrary to our more enlightened judgment. The most uncultivated may be as good in intention as the most learned; although for want of due cultivation, the good seed planted by our Creator, may not have

---

* By these subjects, I mean all those included in my Course of Lectures on Natural Philo-sophy, as well as the present under consideration, as they are all delivered for the same be-neficial purposes to my pupils, whom I am now particularly addressing.

produced that plenitude, which, if the emanation of divine instruction had been shed on their minds, no doubt it would have done.

The unreflecting are likewise objects of pity; then let us not condemn but sympathize with them, and endeavour their reformation, by convincing them of their error, and that in the tenderest manner, not wounding deeper than the case requires; and let our words be like balm to heal that wound.—Let us reflect on the infirmities of human nature—reflect that perfection is not the lot of humanity;—that this life is a state of trial, in which the best of us find many things to combat with; and that all we can do, is to endeavour to fortify our minds against the temptations of sense, by cherishing those ideas most likely to enable us to escape from their allurements; this, which is our duty, will by habit become our inclination, so that we shall derive the highest gratification, as well as the amplest sources of consolation from rational amusements, in whose train follow health, cheerfulness, and joy.

> —— That thou art happy, owe to God;
> That thou continuest such, owe to thyself,
> That is, to thy obedience; therein stand.　　MILTON.

END OF THE THIRD LECTURE.

LECTURE IV.

THE SOLAR SYSTEM EXPLAINED, AND THE NATURE AND AP-
PEARANCES OF THE BODIES WHICH COMPOSE IT PARTICU-
LARLY DISCUSSED.—OF THE QUADRANT AND ITS USES.

THAT most admirable logician Locke, in his Essay on the Human Understanding, very justly observes that—" Though God has
given no innate idea of himself; though he has stamped no original characters on our minds, wherein we may read his Being;
yet having furnished us with those faculties our minds are endowed
with, he has not left himself without witness; since we have sense,
perception, and reason, and therefore cannot want a clear proof of
him as long as we carry ourselves about us."

It is not sufficient to our happiness to know that there is an
all-powerful, all-wise, and ever-enduring Providence; to perfect
our felicity, we wish to discover his attributes—for which desirable
and essential knowledge we need only make use of that reason He
has blessed us with, by applying it to understanding the truths of
christianity, and comparing them with the natural affections of
things, all of which contain the most perfect evidence of the tender mercy, the universal love, and over-ruling providence of the
Almighty. What is it but these assurances that animate us to the
scrutiny of the wonders of his hands, and to the observance of his
laws?—Without them our exertions would become languid, we
should tire in an occupation from which we could derive no ac-
vantage to ourselves or others.—But the reverse being the fact, we
pursue our investigations, both of the natural and spiritual affections of things, with avidity—deriving from them never-failing
sources of delight and satisfaction. Our hearts are humanized and

Plate VI.

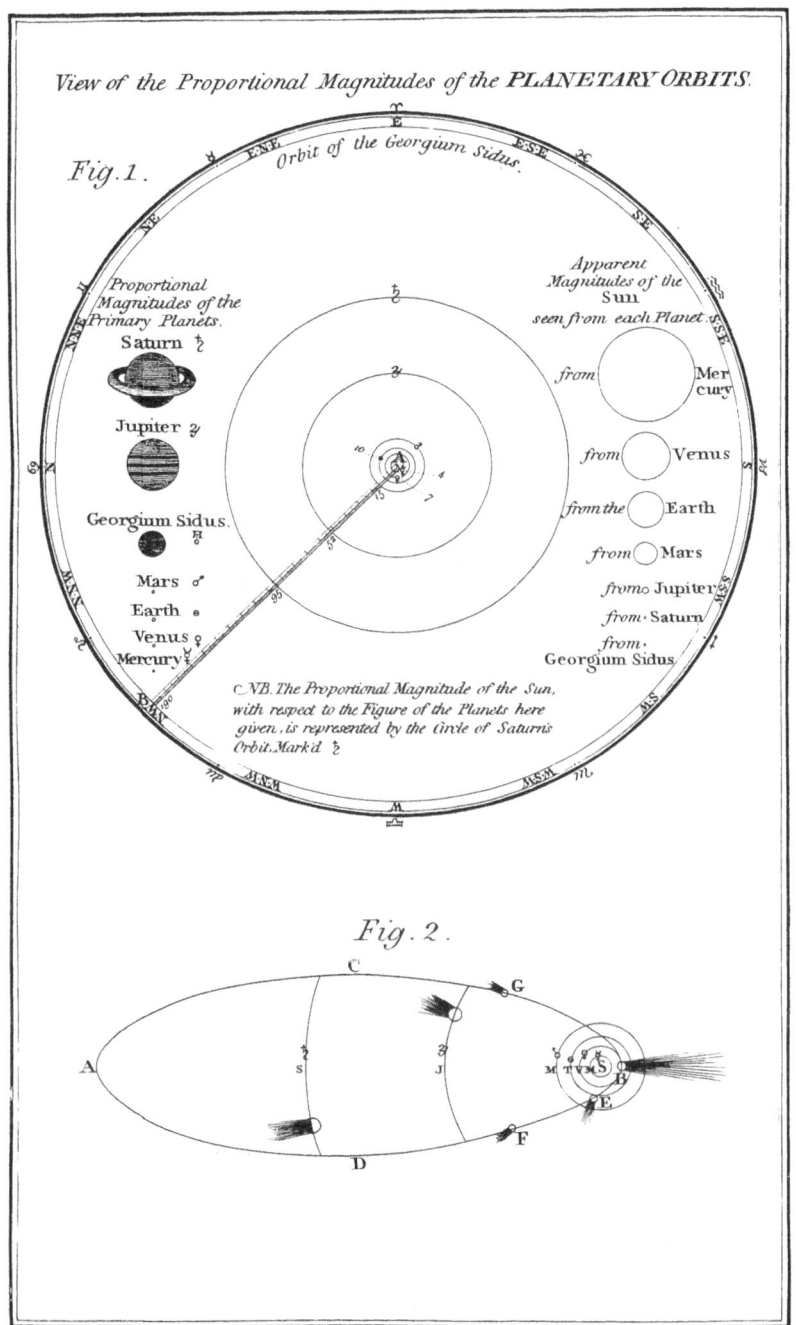

View of the Proportional Magnitudes of the *PLANETARY ORBITS.*

Fig. 1.

Orbit of the Georgium Sidus.

Proportional
Magnitudes of the
Primary Planets.

Saturn ♄

Jupiter ♃

Georgium Sidus.

Mars ♂

Earth ⊕

Venus ♀

Mercury ☿

Apparent
Magnitudes of the
Sun
seen from each Planet.

from — Mercury

from — Venus

from the — Earth

from — Mars

from — Jupiter

from · Saturn

from ·
Georgium Sidus.

C. NB. The Proportional Magnitude of the Sun,
with respect to the Figure of the Planets here
given, is represented by the Circle of Saturn's
Orbit. Mark'd ♄

Fig. 2.

C

G

A

s

J

M
T V
S
B

D

E

F

softened—all gloomy ideas are banished from our minds—we find no more is required of us than what contributes to our happiness, even in this life;—no unnecessary restraints—no austerity of manners—but, on the contrary, that our thankfulness will be best expressed by a cheerful enjoyment of the gifts bestowed on us.

Providence has kindly planted in every human breast a proper sense of right and wrong; therefore if we apply our reason to define the meaning of these intimations, we cannot be misled; but shall find that the service of God is pleasant in the performance, as well as joyful in the end.—For,

Know thou this truth (enough for man to know)
" Virtue alone is happiness below."
The only point where human bliss stands still,
And tastes the good without the fall to ill;
Where only merit constant pay receives,
Is blest in what it takes, and what it gives:
The joy unequal'd if its end it gain,
And if it lose, attended with no pain;
Without satiety, tho' e'er so bless'd,
And but more relish d as the more distress'd.
The broadest mirth unfeeling folly wears,
Less pleasing far than virtue's very tears :
Good from each object, from each place acquir'd,
For ever exercis'd, yet never tir'd ;
Never elated, while one man's oppress'd ;
Never dejected, while another's bless'd ;
And where no wants, no wishes can remain,
Since but to wish more virtue, is to gain.   POPE.

I have informed you that the first thing observed by the ancient Astronomers, was the apparent diurnal motion of the whole heavens; after which, the periodical revolutions of the Sun and Moon excited their attention—to ascertain which required minute observation of the celestial bodies; these led to the discovery of the present system of planets, the peculiar motions of these bodies not escaping the scrutiny of minute observations.

The discovery of the planets did not immediately serve to establish our present system upon just principles, as I have pre-acquainted you; it being thought that the Sun moved round the planets, instead of these moving round him as their centre of motion.

The Solar system, or that which was revived by Copernicus, is now too firmly fixed ever to be doubted, unless the science of Astronomy should again be neglected; and even then, without all writings are expunged, it will not be possible for the present theory to be misconstrued, as the ambiguous legends of the ancients were.

In the Copernican system, which is represented *plate 6*, the Sun is supposed to be at rest in respect of us, in the centre of it, and immoveable; although in fact he has a motion on his axis, that is performed in 25 days 6 hours, which has been discovered by the spots on his disk; as also another motion round the centre of gravity of the whole system, which is occasioned by the various attractions of all the planets.

The Sun is conceived to be seven hundred and sixty-three thousand miles in diameter, and his surface is continually emitting pure light or fire.

The popular opinion of the nature of the Sun is—that it is composed of heterogeneous particles capable of combustion; and not of intirely pure fire; because, when viewed through a telescope, less bright spots are perceived on his surface, the shape of which is variable, as well as the situation and number of them.— These appearances are first obscure, and then become brighter by degrees, so that at last they exceed the other parts of the Sun in brilliancy.—Their obscurity is supposed to be produced by the smoke of the volcanoes, previous to the combustion.

Whatever be the substance of that glorious luminary the Sun, the matter it diffuses is well known, and its uses to mankind strike the mind with sensations like those its influence imparts to all nature—vivifying and expanding.

Those bodies which move round the Sun, regarding him as their centre of motion, are called planets; of which there are now seven discovered. The nearest to the Sun is called Mercury, the next Venus, then our Earth, the next Mars, then Jupiter, then Saturn, and the most remote from the Sun is called the Georgium Sidus or Herschel; all these bodies shine by reflecting the light of the Sun, therefore when they are deprived of his beams, they become totally obscure.

In order to your understanding the construction of the Solar system, and the laws by which it is regulated, I shall endeavour to explain both by figures, accompanied by the theory, preparatory to my illustrating them by the instruments I use; as in doing so, I hope to afford all who may do me the honor of perusing this Treatise, a clear elucidation of these subjects; as well as my pupils an useful theorical retrospect. *Fig.* 1, *plate 6,* represents the situation of the planets in respect to the Sun, and to each other; the character of each being placed on its orbit.

Beside the planets, there are other bodies belonging to our system, called comets, which in one part of their orbit go off too far to be seen by us.

The periods, magnitudes, circumference of the orbits, distances from the Sun, diurnal rotations, and other circumstances of the planets, are exhibited on tables in a subsequent part of these Lectures, containing the representation of the solar system, which I have copied from those of Dr. Hutton, by that gentleman's kind permission.

It is well known, that all the other planets, (as well as our Earth) of which we are able to discover the spots on their surfaces, have a motion on their axis—by the regular change of situation observed of those spots,—from which we may infer, that those of which we cannot perceive these effects have the same.

By the irregular appearance of the motions of the planets in

M

their orbits, as seen from our Earth, sometimes moving backward or retrograde, at others appearing stationary, at others moving direct, being the result of their having a progressive motion round the Sun, like that of our Earth, we do not hesitate to acknowledge these clear evidences of their doing so, and also that they respect the same central body as the centre of their motion.

Perceiving the planets to be regulated by the same laws as our Earth, we naturally conclude that they also are the residences of rational creatures, like ourselves, endowed with the like sensibilities, capable of enjoying an equal portion of the gifts of Providence, as they receive an equal share of heavenly influence.

Mercury is the nearest planet to the Sun, being only thirty-seven million of miles distant from him, which, compared with the remoteness of the other planets, may be termed near, although in fact it is an immense distance; yet such is the density of the Sun's light at that distance from him, that Mercury is seldom seen, being lost in the splendor of his rays, by which we are prevented from ascertaining this planet's motion on its axis, no spots being discoverable, so highly is its surface illuminated.

As Mercury is never so far removed from the Sun as Venus, or the other planets of our system, we know that his orbit must be included within those of all the rest; another circumstance corroborates this opinion:—when a number of bodies move round another as their centre of motion, the squares of their periodic times are proportional to the cubes of their distances from the central body *.

Mercury performs his revolution round the Sun in about 87 days 23 hours, which is a much shorter time than any other planet per-

* Any number which is multiplied by itself is called the square of that number; as suppose 2 be multiplied by itself, the square of that number is 4, and 2 is called the root of 4, or the square root. If a square number be multiplied by its root, as 4 by 2, the product 8 is called the cube, or third power of 2, which, in respect to the 8, is called the cube root.

forms its annual revolution in, the length of his year being not quite three months of ours. His daily motion is not known, for the reasons before given. His diameter is supposed to be equal to three thousand two hundred miles; but, on account of his nearness to the Sun, his motion is so quick, as not to admit of accurate mensuration, being at the mean rate of one hundred and five thousand miles an hour.

Mercury exhibits phases like the Moon, which proves that he shines only by reflected light, notwithstanding his peculiar brilliancy. The times for the observation of this planet are a little after sun set, and a little before sun rise.

Venus, the next in order, appears the most beautiful star of the Heavens, as viewed from our Earth; she is denominated the morning and evening star, and is the next planet to Mercury in our system. Her distance from the Sun is computed at sixty-eight million of miles, and her mean motion in her orbit at the rate of seventy-six thousand miles an hour. The circumference of her orbit being four hundred and thirty-three million of miles, and her annual revolution being performed in 224 of our days and 17 hours, which is the length of her year, not quite two-thirds of ours.

The diameter of Venus is computed at seven thousand seven hundred English miles; her rotation on her axis is performed in 23 of our hours and 22 minutes.

That the Earth is not within the orbits of Venus and Mercury, is evident, by these planets never being visible when in conjunction with the Sun, or in opposition to that luminary, i. e. we never see these stars shine when they are in the same sign with the Sun, or in the opposite sign to it.—When Venus is in the same sign with the Sun, and directly in a line between us and that luminary, she appears only a dark spot on the Sun's disk, and when she passes over the disk of the Sun, her passage over it is called the transit of Venus.

Venus and Mercury are called inferior planets, because their orbits are included within that of the Earth. These planets are never seen in the eastern point of the heavens when the Sun is in the western, on account of their near approach to that luminary — following close after him in the evening, when he retires from our view — and immediately preceding him in the morning, announcing his returning influence, which gladdens the face of nature by dissipating the incumbent gloom.

Whilst Venus is performing her apparent office of attendance, she is called Hesperus, or Vesper, the evening star, in which she continues 290 days ;—when that of Harbinger, announcing the Sun's approach to gladden the Earth, Phosphorus, the morning star ; in which office she remains the same time as in the former.

This planet exhibits phases like the Moon ; and spots similar to those perceived on that satellite have been discovered on her surface, by which her rotation on her axis has been ascertained.

Our Earth is the next planet in the system. Of this planet I have already said much, and shall say a great deal more ; but in placing it in its situation in regard to the other planets, I shall only treat of it in those respects in which I have the others. Its diameter is feven thousand nine hundred miles, the circumference of its orbit five hundred ninety-eight millions, and its mean annual motion fifty-eight thousand miles an hour, by which it performs its period in 365 days, 5 hours and 49 minutes. Its mean distance from the Sun is ninety-five million of miles, therefore its motion in its orbit is not much more than half as swift as that of Mercury in his. The Earth is accompanied in its annual progress by one Moon, for the most beneficial purposes.

Mars is the next planet in the system, and his orbit is consequently exterior to that of our Earth, his motion in which is computed at a mean rate fifty-five thousand miles an hour ; its circumference being nine hundred and twelve million of miles, through which he revolves in 686 of our days and 23 hours. His diameter is near four thousand two hundred miles; his distance from the Sun one hundred and forty-four million of miles; and he performs his diurnal rotation in 24 of our hours and 39 minutes. The light reflected from this planet appears of a reddish dusky colour; which appearance, added to the deficiency of reflected light (this planet not having a Moon) gives the idea of its atmosphere being more dense than ours, and that thus, by its greater refractive power, it supplies the place of a Moon to that planet, refracting and reflecting the light of the Sun for a considerable length of time before that luminary rises above the horizon of places on that globe, and after he has set below it.—That excellent astronomer, Dr. Herschel, has discovered two white circles round the poles of this planet, which he thinks originate from the snow lying about those parts. The diurnal rotation of this planet is discoverable by the spots on its surface.

Mars, when in that part of his orbit in which the Sun is between us and him, is five times further removed from our Earth than when our Earth is interposed between the Sun and that planet. In the former situation he is said to be in conjunction, and when in the latter in opposition, in reference to his situation in respect to the Sun as seen from our Earth. According to his distance from us and the Sun at any time, such will be and appear his illumination; and according to the former his apparent size. In the quadratures he appears like a half-moon, but never horned like Venus, which convinces us that his orbit is exterior to ours. This you will perfectly understand by the planetarium. This planet being near twice the distance from the Sun that our Earth is, his diameter, as seen by the inhabitants of Mars, must appear but about half as large as it does to us; and the heat and light be not more than one-fourth* part so great as on our globe; their year is near twice as long as ours.

* Heat diminishing in proportion to the squares of the distances, and the apparent diameters of objects in proportion to the distances themselves inversely.

It is perceived by the revolution of the spots on the disk of Mars, that his axis is nearly at right angles to   plane of his orbit; which must cause the days and nights to be nearly equal to every part of this globe.

Jupiter, the next planet in the system, affords great scope for speculation, many things being observable of it not exhibited by any other planet of the system:—such as the variety of its lustre on different parts of its surface, some parts appearing of a duller and fainter illumination, and these latter occupying spaces on the planet similar to zones or belts, that run parallel to each other and to the plane of its orbit, to which its axis inclines but very little indeed; on which account the days and nights must be nearly equal to each other; and all over this globe it is the same as with Mars, a perpetual equinox.

The belts are not always discoverable on the face of Jupiter, sometimes eight of them are perceived, sometimes not more than one; if the changes in their appearances are observed, they seem to be occasioned by the flowing of one belt into another—as while one is decreasing another is increasing; and what further corroborates this opinion is, there is lying between the belts obliquely, at the time of these changes, a stratum of the same hue with themselves.

Bright spots also appear on the face of this planet, one of which is so remarkable, that the time of the rotation of Jupiter on his axis is ascertained by it.

After the discovery of this remarkable and stationary spot on the face of Jupiter, it disappeared, and did not re-appear till fourteen years afterwards, and then it was exactly in the same place, where also it has continued to appear every fourteenth year since the discovery, which was made in 1694, by which his period is also determined.

This planet is the largest in the system, his diameter being computed at eighty-nine thousand miles, which is eleven times as much as that of our Earth; and yet such is the amazing velocity of his motion on his axis, that he revolves round it in 9 hours 56 minutes. The circumference of his orbit is estimated at three thousand one hundred million of miles, and his motion in it twenty-nine thousand miles an hour; so that he performs his annual revolution in 4332 days 9 hours. His mean distance from the Sun is four hundred and ninety million of miles. Jupiter is attended constantly by four Moons, which revolve round him, or rather round the centre of gravity between the planet and themselves—these must afford Jupiter a great portion of reflected light.

This planet is also nearer to our Earth when in opposition, than when in conjunction with the Sun; and his mean distance from that luminary being a little more than five times further than that of our Earth, the inhabitants of that planet must feel but one twenty-fifth part of the influences of the Sun felt by us; and to them his apparent diameter must be only one-fifth part of what it appears to us.

The four satellites of Jupiter occasion that planet to have four different months, according to the period of each. It is supposed that the shortest of these months contains only $4\frac{1}{4}$ days, and the longest 41 days; the second about $8\frac{1}{2}$, and the third $17\frac{1}{4}$ days.

The distances of Jupiter's satellites from each other render them very distinguishable, and the shadow they cast on his disk when interposed between him and the Sun, as well as their eclipses in passing through his shadow, prove them to be opake bodies, and that they serve to afford reflected light to Jupiter in the absence of the Sun, as the Moon of our Earth affords us.

The discovery of the satellites of Jupiter has proved of some aid to navigators, the eclipses of which have afforded them the opportunity of computing their longitude at sea. As thus—having noted the time in White's Ephemeris, or the Nautical Almanac, in which a satellite will be immersed in the shadow of its primary, in the longitude of Greenwich, suppose it is at 7 o'clock in the evening of such a day; and being at sea, if we do not perceive this circumstance till 9 o'clock at night, the difference of time being two hours, by allowing the number of degrees of the equator equal to that space, which is known to be 15 degrees to an hour, we discover that we are in 30 degrees of longitude, which is east, because the time is later than at London or Greenwich,—if it happens two hours sooner, then we know that we are so many degrees west of London or Greenwich. In practice the observation is taken with greater accuracy, allowing for the difference of time-pieces; but the motion of our Earth on its axis being uniformly performed in the same time, 15 degrees of the equator or circle of our globe, directly situated between the two poles of the axis of motion, must always be equal to one twenty-fourth part of the time in which it performs its revolution.

The velocity of light, I have previously informed you, was also ascertained by these eclipses; so that we may readily admit that many important truths have been sanctioned by this admirable investigation, for the idea of which we are indebted to Gallileo.

Saturn, the next in the system beyond Jupiter, is so far removed from us and the Sun, that investigations of his phœnomena are not so attainable as those of Mars or Jupiter. His motion on his axis has not yet been ascertained, as we cannot perceive sufficient of the spots on his surface to estimate it by.

His distance from the Sun is nine hundred million of miles; and he performs his annual revolution in rather more than 29 of our years. His diameter is supposed to be seventy-nine thousand miles. He is encircled by a luminous ring, which is nearly as bright as the planet, which phœnomenon has occasioned various conjectures, but have produced nothing decisive. It appears detached from the planet, as in some situations a star has been seen between

the ring and its primary; for such I conceive it most natural to call
the planet, on whose affections it must I think be dependant; as on
no other grounds can we account for its constant attendance on it,
or the changes of its appearances; although some philosophers
have formed a different idea of their connection, in the infancy of
the observation, which subsequent observations seem to invalidate:
they have supposed this white appearance exterior to the planet to
be produced by the vapours which issue from its surface, forming a
sort of continuous white cloud;—others, that it is occasioned by
a vast number of satellites disposed in the same plane, which by
their affinity blend their lights together, and so form the appearance
of one luminous body, which affords an abundant quantity of reflected light to that remote planet of the Solar system. This, if
necessary, would consistently account for it; but Saturn being attended by seven moons, precludes that idea, as those must be
quite sufficient to make up for the deficiency of intrinsic light, at
the distance of those Moons from the Sun, by which their illumination must be fainter than that of the Moon which illumines our
nights.

Yet as nothing is created in vain, we naturally wish to find out
the use of this extraordinary light, which we may suppose essential to the comfort of the inhabitants of that planet; which desire
prompts me to offer my conjecture on that subject—and will, I
trust, be an excuse for my temerity.

On account of the immense distance of the Sun from this planet,
may it not be necessary that it should be encompassed by a peculiar fluid mass, exterior to its atmosphere, to collect the rays of
light before they fall into the region of the atmosphere, by which
they may acquire additional brilliancy and force; and thus com-
pensate to the inhabitants for the distance at which they are placed
from their grand vivifying principle, by transmitting and reflecting
the light, with such due lustre and heat as may be essential to the

N

circumstances of that planet?—This opinion I offer with extreme diffidence, as probably, in endeavouring to throw additional light on the subject, I may only evince the necessity of additional light to my understanding, in order to supply the deficiency of my judgment.

When our eye is in the plane of Saturn's ring, it becomes almost invisible to us, appearing only as a dark band or line across the body of the planet. It disappears twice in every revolution of that planet round the Sun; at other times it appears more or less inclined to the planet, and sometimes surrounds the outward circle of its illuminated face, so that we can distinguish it clearly from the planet. When the eye of an observer is raised above the plane of the ring, a shadow is perceived above it on the planet.—Besides this ring, Saturn has two luminous belts, which are permanent.

Saturn is about one thousand times as big as our Earth, and three-quarters the size of Jupiter. He always shows a full round face, but his light appears dull to us on account of his immense distance; from which latter circumstance his revolution on his axis cannot be ascertained by an observer on our Earth. His Moons all move round his equator, which is inclined to the plane of his orbit in about 30 degrees.

Saturn's ring appears to us quite distinct from his disk when the planet is in 20 degrees of Sagitarius, and then the northern parts of the planet are turned towards the Sun, when it must be summer to those places situated north of his equator.

When he is in 20 degrees of Pisces the ring is quite shut, appearing only as a line upon the equator; and the equator being turned towards the Sun, the days and nights must be equal to all the inhabitants of that planet at that time.—When the planet is in 20 degrees of Gemini, the southern parts of it are turned towards the Sun, and the ring again appears luminous and distinct; this is the time of summer with the inhabitants of those parts.—When he arrives at Libra, the Sun again being over his equator, makes the days and nights equal, and the ring again appears only like a line across the planet.

These positions seem to sanction my idea of the ring being a peculiar region, adapted to the purpose of affecting the intrinsic light in its approach to this planet; as it does not appear luminous but in those parts which receive the direct influence of the Sun's beams.

The last discovered planet, and most remote in the system, is the Georgium Sidus, which was discovered by Dr. Herschel. It is not disernable without the aid of glasses of considerable power, on account of its immense distance from the Sun and from our Earth. It was first discovered near one of the feet of Gemini. Its distance from the Sun being twice that of Saturn, many of its phœnomena have not been discovered.—Only two Moons have as yet been perceived to attend on this planet, though doubtless, on account of its remote situation in respect to the Sun, it has more. The regularity of its motion corresponding to that of the other planets, and its moving in the same direction, prove it to be one of our system. Its mean distance from the Sun, which has been calculated from the observation of it when at its nearest and greatest distance from that luminary, is nineteen times greater than that of our Earth from it. By observation of its progress in different parts of its orbit, it has been found that it moves through it in about 82 years, and that it is ninety times larger than our Earth.

From the different phases they exhibit, it is evident that all the satellites of the planets are opake bodies, shining only by reflecting the light of the Sun; and that they revolve round their primaries as those revolve round that grand vivifying principle, situated in the centre of all their motions by the kind and all-wise ordination of Providence, by which they are compelled to accompany their respective planets through the whole of their annual revolutions.

Mars, Jupiter, Saturn, and the Georgium Sidus, include the orbit of our Earth, which is made evident to us by their elongations and different appearances, as I shall hereafter illustrate. Also, that the Sun is the central body round which they must revolve, and that their orbits are situated in respect to each other as I have described.

The comets of our system remain to be explained; the astronomy of which, although imperfect, approaches nearer to certainty than that of those who, living in remote ages, were unacquainted with the laws of gravity, and of those also which regulate the subtile effluvia of the Sun, &c.

These extraordinary bodies are found, by their reflective power, to be opake.—The matter of heat and light darts from them like fiery tails;—as when an insulated jar is receiving a full charge from the electrical machine, it throws off its redundancy; so do the comets emit a stream of fire from their bodies on the side opposed to the Sun, from which they receive their super-abundant fire;—therefore, if I may be allowed to reason from analogy, I should suppose that by this mode do the comets throw off the redundant heat which they must receive from the Sun, at the times of their nearest approaches to him; and conscious of the happy equilibrium sought after by all natural bodies, which is happily attainable by the constitutional construction of their parts, I venture to intimate the possibility of this being the cause of the effect we perceive of these motions called fiery tails;—that the additional heat thus issuing from these bodies, prevents an accumulation unfavorable to animal existence, supposing these bodies to be inhabited, as we have great reason to do, seeing that God has created nothing in vain.

If my opinion of the appearances I have been adverting to is futile, or any arguments can be produced to invalidate it, I am open to conviction; but till then I shall retain it, and think myself excusable in advancing my ideas on the subject, till the physical causes of these appearances can be ascertained—all that has hitherto been said of them, like my observation, amounting to no more than mere hypothesis.

The other circumstances of the comets have been pretty accurately ascertained by the admirable and sublime Newton, such as the degrees of heat, light, and motion, communicated to them. Estimating the degree of heat each must receive from the Sun on its nearest approach to that luminary, he supposed the one seen in 1680 might have retained its heat for twenty thousand years, supposing it were necessary that it should, which we may conclude not to be the case, and that therefore such an accumulation never actually takes place. In this supposition I do not presumptuously mean to confute his assertion, as I conceive that great man by no means intended to convey an erroneous opinion, but that he meant merely to shew us the immensity of that power which could suffer such extremes of heat and cold, without altering the happy indispensible harmony of nature; and I am the more assured of this, as although the heat he estimated might be received by the comet which appeared in 1680, was so intense and profuse, as by his calculation it would have been sufficient for the animation of that planet for twenty thousand years; yet as he immediately after discovered that its period was not more than five hundred and seventy-five years, we may conclude so great a naturalist as this great man, could not admit that the comet actually did retain so superabundant a quantity of heat; therefore I conceive that I am justified in asserting, that such an inference not being deducible from his calculation of the period of this comet, or from the universal laws of nature, although he advanced the postulatum, he meant not to convey such an idea, as that they actually would retain such a superabundant quantity of heat.—But it is unnecessary for me to perplex you with speculations which all nature so clearly determines upon.

The comets all revolve round the Sun in elliptical orbits; but the cause why all the bodies of our system do so, I shall inform you of, (when treating more particularly of their periods,) and why those of the comets are peculiarly long ellipses.

These bodies disappear from our view, and again become visible, according to their situation in respect to the orbit of our Earth; as in revolving in very long ellipses, they are sometimes too remote for our inspection, their greatest distance from the Sun being far beyond the orbit of the Georgium Sidus, as these bodies are not much larger than our Moon. When they are in those parts of their orbits in which we can see them, some things may be observed of them, which seem to corroborate my ideas.

That the tail or fiery matter exhaled from these bodies is always in the direction of the Sun's beams, and on the opposite side of the comet from the Sun; and the nearer they are to that luminary, the longer is the stream of fire emitted from them. This luminous tract appears differently formed, according to the situation of it in respect of an observer on our Earth.

In *fig. 2, plate 6,* let S represent our Sun, *M* the orbit of Mercury, V that of Venus, T that of our Earth, M of Mars, J of Jupiter, S of Saturn, and A B C D that of a comet; A at its greatest distance from the Sun, and B its least. You observe, that in the former it is far beyond all the orbits of the planets, and in the latter it is almost as near to the Sun as within one-sixth part of the Sun's diameter from it. When the comet arrives at F, which is near the orbit of Mars, it emits but a small quantity of fire; as it approaches nearer to the Sun, as at E, the tail becomes longer; when at B it appears longest of all, being its nearest situation to the Sun; as it revolves from B to D, receding from the Sun by degrees, the tail becomes less and less; and when it arrives at G, the orbit of Jupiter, it is not seen without a telescope of great power. The length of this tail is sometimes as much as 40 degrees of the Heavens, a space equal to about eighty million miles. The periods of the comets, although they may be, and have been pretty accurately ascertained, yet on account of the length of that part of their orbits, in which they are hid from our view, and the time taken up to perform it in, also their similar appearances, no great accuracy in this part of Astronomy can be expected, as their motion in some parts of their orbits may be slower than is calculated for; and when we do see a comet, we can no otherwise assert its identity, that it is the one expected to appear at the time, but by comparing its direction through the great expanse, with the track pursued in the other case; in which latter also we are subject to be deceived, as there are an innumerable multitude of comets, and the orbits of some of them may be performed so nearly in the same plane, as to baffle our discrimination.

That their periods have not been fixed upon immutable principles, is evident, by those which, having been expected at stated times, at one time appearing agreeably to the prediction, at another within a certain short space of time after the period previously anticipated, and at another not at all. The two former circumstances were observed of the one which was first seen in 1456, the time of which was calculated at $75\frac{1}{2}$ years, and which was actually seen, or supposed to be seen, as a comet did appear at the several periods, i. e. in 1531, 1607, and 1682; but in 1758, when it was again expected, it did not appear; but one was visible in 1759, about six months after the predicted time, which was supposed to be the same. The latter circumstance, of a comet appearing, as supposed, exactly at the period fixed, at one time, and not appearing again at or near its expected return, was experienced respecting the one which was seen in 1661, which being 129 years from its former appearance, occasioned great expectation of its return in 1789, but no comet then appeared, or has since been perceived.

These bodies revolve round the Sun in all directions, some moving easterly, some westerly, some northerly, some southerly; some obliquely, and some perpendicularly, to the plane of the planets orbits. These circumstances must render it still more difficult to ascertain their identity and affections;—indeed, I think the greatest philosophers and ablest mathematicians, must confess that the nature and affections of these bodies are far beyond their scrutiny and calculations.

Three of the comets have had their periods settled, as near the truth as could be expected, yet not, I think, permanently so. The first of these is that which appeared in 1532, 1607, 1682, 1759, which I have before mentioned; the latter appearance of which did not exactly answer to the foregoing periods;—should it in future conform to either, its period being supposed about 75 years, we may expect its return in 1834.

The second appeared in 1532, 1661, and was expected in 1789, but has not yet appeared; therefore we cannot suppose the period of this yet fixed.

The third is the one Sir Isaac Newton estimated the heat of in 1680, and the period of which he settled at 575 years; the accuracy of which calculation cannot be decided till the year 2255. The velocity of the latter comet's motion, when it was nearest to the Sun, was computed at eight hundred and eighty thousand miles an hour.

When these motions were first discovered, it was supposed they were the repositories of the wicked, where they received punishment for their crimes, and that their torments were occasioned by the extremes of heat and cold; which ideas originated in the opinion of the impossibility of animal existence being supported under such vicissitudes; but thanks to science and investigation, these dreadful ideas are removed, we can view these glories of the Heavens without horror; and though we are not certain of their being made for human beings in their rational state, yet we are conscious that God can accommodate the animal body to any degree of heat or cold, if he pleases; supposing that these extremes are experienced by the inhabitants of those bodies, which I do not think is the case, as I imagine that the superabundant heat is thrown off from those globes in the manner I have been describing; and only a sufficient quantity retained for all the circumstances of these planets.

> Let stupid Atheists boast the Atomic dance,
> And call these beauteous worlds the work of chance;
> But nobler minds, from guilt and passion free,
> Where truth unclouded darts her heav nly ray,
> Or on the Earth, or in the etherial road,
> Surveys the footsteps of a ruling God;
> Whose single fiat form'd the amazing whole,
> And taught the new-born planets where to roll;
> With wise direction curv'd their steady course,
> Imprest the central and projectile force,
> Lest in one mass their orbs confus'd should run,
> Drawn by the attractive virtue of the Sun,
> Or quit the harmonious round and wildly stray
> Beyond the limits of his genial ray.   Mrs. Eliz. Carter.

Having dwelt sufficiently on the circumstances of the size, distances, &c. of the planets of our system, it must naturally have excited your curiosity to know how these have been ascertained of objects so remote, and that are not accessible.

I shall therefore endeavour to convey to you some idea of the mode by which mathematicians acquire their knowledge of the comparative size and distances of the heavenly bodies of our system, and of their distances from each other; by the application of which observation to the science of mixed mathematics, their real size and distances are known. The instrument thus employed is called a

O

quadrant; by which the relative size and distances of the bodies of our system are ascertained, and a just idea of their comparative motions is formed, which renders it extensively useful in practical Astronomy.

Every circle, great or small, being by mathematicians divided into 360 parts, which they call degrees; * ninety degrees, or a quarter of a circle, is sufficient to measure all angles, formed between the horizon of any place on our Earth, and the line perpendicular to it, which goes up to the zenith, as *fig.* 1, *plate* 2, where the line BC represents the plane of the horizon, and ABC the quadrant, AC the perpendicular to the horizon, and A the zenith point.

An angle is the opening between any two lines which touch each other in a point; and the width of the opening determines the portion of the circle included within the angle, the length of the lines not being in the least considered.

Let the lines CA and CB, *fig.* 1, *plate* 7, represent a pair of compasses, and the curved lines AB, DE, and FG, the quarter of so many circles of different sizes. You perceive that although each of these differs from the others in size, yet that each contains the same portion of, or a quarter of a circle; and thus it would be from the smallest to the largest that could be formed, that they would all contain exactly 90 degrees each; by which simple observation the comparative measure of angles may be extended to an indefinite distance.

The altitude of a star above our horizon, and its situation in respect to the points of the compass, may also be known by the application of this instrument; the one represented *fig.* 2, *plate* 7, is called a quadrant of altitude, from that use of it.

It is mounted on a pedestal, and is moveable round an axis at

---

* A degree, when used in relation to the circumference of our Earth, includes a space of 69 English miles.

D, which affords the opportunity of placing the quadrant in any vertical position, inclining it more or less to the horizon, as occasion requires.

The circle EF of this instrument is intended to represent the plane of the horizon; and on it the points of the compass are graduated.

In the centre of this plane, the pedestal moves in a socket, so that the quadrant may be directed towards any point of the horizon. Upon the radius CB of the quadrant, two sights are perforated; on the point B a dark glass is placed, to use for the inspection of the Sun. The end C has a large hole, with cross wires, by which we may perceive the ascent or descent of the Sun, or any other object, through the opposite sight. There is also below it a small hole, which is used to take the Sun's altitude on the opposite sight. The arm E shews the point of the compass in which a celestial object appears.

The screws, G H I, are to adjust the plane of the quadrant to the exact level of the horizon, which is the first part of the adjustment of this instrument, and which is ascertained by placing the plane of the quadrant in different situations of it, in respect to the points of the compass, till the line of sight is in all of them perpendicular to the plate of the quadrant, or plane of the horizon, which may be effected by placing the line of sight only in two opposite situations of the horizon, and observing that the plumb-line touches the plane of the quadrant in both of them. Thus—

Bring the index to 0°, and mark the degree cut by the plumb-line on the plane of the quadrant; then move the index to 180°, which is exactly opposite to 0°, and observe the degree cut by the plumb-line; if it is the same as in the former adjustment, then there will be no occasion to alter the screws; but if it is not, one or both of the screws between 60° and 70° must be turned, till the plumb-line intersects the middle degree, or part between the two foregoing degrees marked; then observing what that degree is, the index must be turned to both the 90°, being at right angles to the former part of the operation; and the adjustment of the plumb-line to the last observation, or middle degree, must be made by turning the screw at 180°, observing not to alter the other screws. After one or both of these adjustments, the quadrant will be ready for use, and will shew you the distance of any celestial object from the zenith point, that is, how much it wants of being perpendicular to you; also its height above your horizon. Having found the object, direct the line of sight towards it; and count the number of degrees from the plumb-line to the end of the quadrant next your eye, and that will shew the distance of the object from the zenith; and the degrees on the quadrant from the plumb-line, to the end of the quadrant furthest from your eye, will shew you its height or altitude above your horizon. Lastly, to know what point of the compass it is on, look at the degree on the plate of the horizon, to which the quadrant is directed, allowing for the variation of the compass, which you should always fix the horizon of the quadrant to, previous to making your observation, which is now 23° $\frac{1}{2}$ west at London.

By a quadrant, and similar instruments, the principles of which are so simple that any person may understand them, we are able to obtain an accurate relative measurement of the circles of the Heavens, and the distances of the heavenly bodies from each other, which enable us to authenticate the observations of the learned, although we are not able to pursue their mode of investigation, in ascertaining the real sizes and distances of those sublime objects; and thus we reap the advantage of abler heads, without the trouble and application by which they procure that information they so liberally afford us. Surely we are much indebted to those great men, who by unwearied assiduity have supplied us with such useful knowledge, whereby we are enabled to apply their calculations without any great mental exertion, and are as certain of their validity as if we were to fathom the depth of that sublime science which furnishes us with such valuable truths, by the observation of those circumstances which corroborate their previous intimations.

Modern Astronomers and Mathematicians deserve our highest esteem and commendation, who, by a communicable spirit, liberally extend the province of science, and thereby confer a lasting benefit on society; not like the ancients, who, contracted in their ideas, and wishing to be considered as superior beings, hoarded up their scientific principles for the purpose of ostentation;—and what was the consequence of their selfish and mysterious conduct? —Nothing less than in some cases the total subversion of all their knowledge, which, dying with them, the succeeding generations, uninstructed in their theories and investigations, and perhaps not possessing the same spirit of enquiry, the science became neglected or perverted; and thus all the advantages which had been procured were lost to mankind. —Such conduct is like the folly of the avaricious man, who hoards up that which at his death cannot profit him; and which devolving to those who have been circumscribed by his parsimony, are unacquainted with the use of riches, set no value upon the means they possess of communicable kindness, and rational amusements; too frequently run into excesses, and squander their time in useless if not dangerous pursuits, by which the intention of their predecessor is frustrated—Whereas, when both the philosopher and the rich man give glory to God by a free communication of their gifts, and a proper application of them, the science of the former is extended, and the riches of the latter usefully employed.

What Armstrong has pointed out as the proper use of riches, is equally applicable to the extensive communication of intellectual acquirements; and therefore, in agreement with my comparison between them, I shall conclude this Lecture in his words:

> But for one end, one much neglected use,
> Are riches worth your care (for Nature's wants
> Are few, and without opulence supply'd):
> This noble end is, to produce the soul;
> To shew the virtues in their fairest light;
> To make humanity the minister
> Of bounteous Providence; and teach the breast
> That gen rous luxury the gods enjoy.   ARMSTRONG.

END OF THE FOURTH LECTURE.

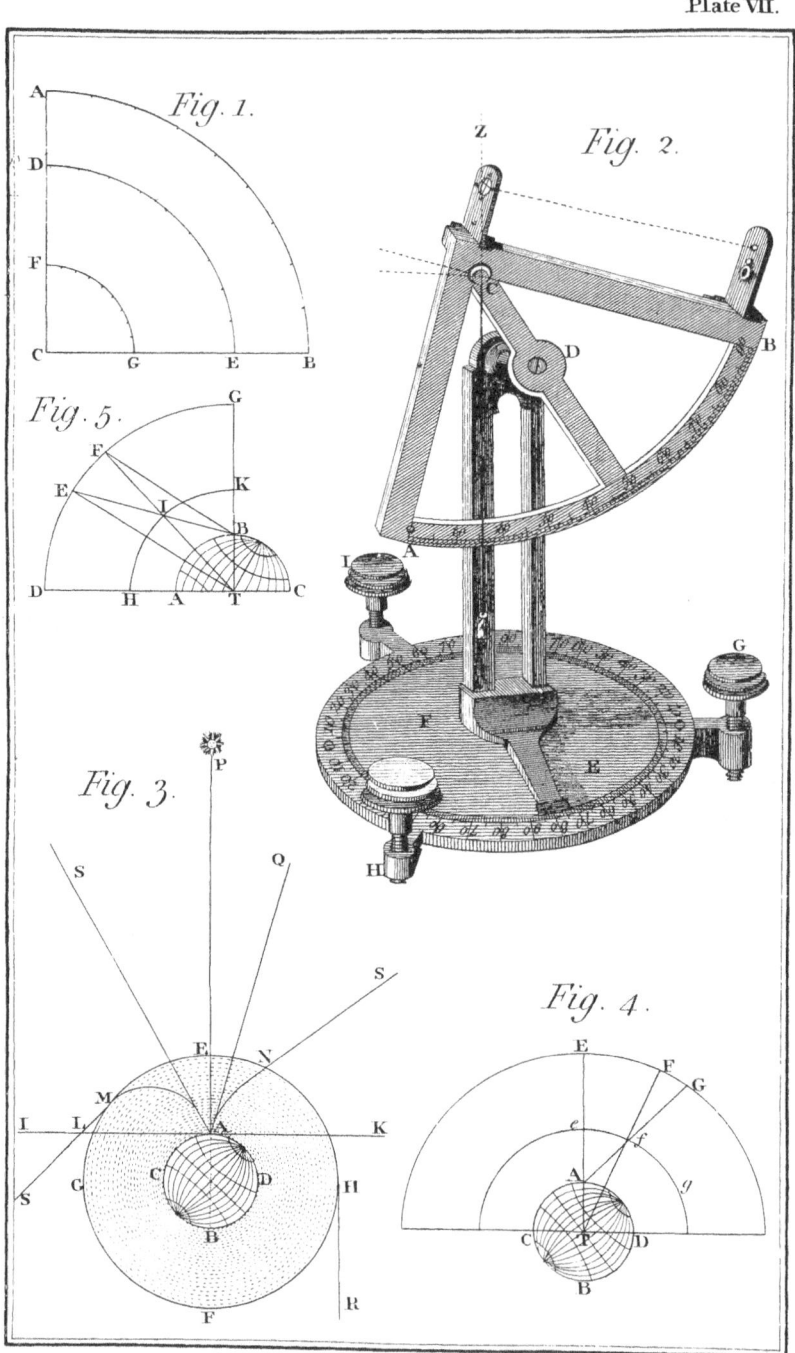

Plate VII.

Fig. 1.

Fig. 5.

Fig. 2.

Fig. 3.

Fig. 4.

LECTURE V.

THE MANNER IN WHICH ASTRONOMERS AND MATHEMATICIANS
ESTIMATE THE REAL SIZES AND DISTANCES OF THE BO-
DIES BELONGING TO THE SOLAR SYSTEM.——OF THE
ABERRATION OF LIGHT, AND HOW IT IS OBTAINED.

> Of heav'n my strains begun ; from heav'n descends
> The flame of genius to the human breast.
> But not alike to ev'ry mortal eye
> Is this great scene unveil'd : For since the claims.
> Of social life, to diff'rent labours urge
> The active powers of man ; with wise intent
> The hand of nature, on peculiar minds,
> Imprints a different bias, and to each
> Decrees its province in the common toil.   AKENSIDE.

MATHEMATICIANS and Astronomers acquire their knowledge of the real sizes, distances, and situations of the heavenly bodies of our system by means of plane and spherical trigonometry, paralax and refraction—the apparent diameters of bodies at certain distances from our Earth—the diminution and velocity of light ; investigations minutely and deeply entered into by them ; but which studies, extensively considered, do not come within the sphere of knowledge I wish to conduct my pupils into, being unnecessary to their peculiar province in life to comprehend all the phœnomena they prove ; more especially as we are furnished with the result of them all, which we are enabled to apply to our intire satisfaction. The study of mathematics would be a misapplication of your time, which might be justly attributed to vanity and ostentation, and be considered unbecoming your character as females, by employing that time which is more usefully occupied in pursuits adapted to your situation in society, and as the validity of astronomical computation may be proved by those instruments I have provided, aided by your reason.

For the foregoing reasons I shall also avoid introducing into this work astronomical tables for computation, although I might do so from those authors who have published them; but as you would not understand the principles upon which they are constructed, they could afford you no more satisfaction than you can derive from merely examining the result of them in an Ephemeris, as being unacquainted with the sublimer branches of the mathematics, you could only acquire a superficial knowledge of them. Nor could I, for the same reasons, define them properly; and therefore to attempt it would be deviating from my settled rule of conduct in all I instruct you in, by communicating that which I did not fully comprehend—a folly of all others the most inexcusable in those who teach, and one which I trust will never be imputed to me.

To those who, being acquainted with the elements of mathematics, wish to understand the construction of Astronomical Tables, I recommend Dr. Hutton's Mathematical Dictionary, in which they will find ample scope for their abilities, as well as the most clear elucidations, which are calculated to instruct, not confound, the student.

Having premised so much, I shall content myself with bringing you acquainted with the nature of those things mathematicians apply to the sciences they cultivate, in order to ascertain the real sizes, distances, &c. of the heavenly bodies, without perplexing you with theories too profound for your contemplation.

What I have already related of the general nature and properties of light, will render the postulatums I am about advancing intelligible; and the theorems in trigonometry subjoined to these

Lectures will assist you in comprehending the practical part of the subject.

When light falls obliquely on a medium of a different density, you know it is refracted; you are also acquainted with the laws of refraction, and that heat and light diminish as the squares of the distances increase from the object which emits them; also that the apparent diameters of bodies decrease in proportion as we are further removed from them, directly as the distances themselves; i. e. at twice a given distance they appear half as large, and at four times that distance a quarter as large, and at five times the distance only a fifth part as large, and so on. But heat and light diminish as the squares of the distances, as thus: At twice the distance they are four times less hot and light; at three times the distance nine times less, and so on. The bulk or solid contents of the heavenly bodies are computed by their diameters, and their comparative bulks by the cubes of their diameters; i. e. if the diameter of one body be double that of another, then its bulk is eight times larger than that of the smaller; if the diameter be three times greater, then it is twenty-seven times bigger, being calculated thus: twice 2 is 4, and twice 4 is 8—also, 3 times 3 is 9, and 3 times 9 is 27. Let us now apply these general laws to particulars.

As the quantity of heat and light thrown off from the surface of the Sun, must become thinner and thinner as it recedes from him, on account of its divergence; those planets of our system, which are most remote from that luminary, must have a less portion of his influence impressed on them than those which are situated nearer to him; and this force must be inversely in proportion to the squares of their distances, throughout the whole system of planets; which positions the observations of Astronomers upon the affections of the planets serve to authenticate.

The direct influence of the Sun's rays can never be received on any part of our globe but that to which he is vertical, so that he

P

is never seen in his right place by any but those to whom he appears in their zenith point; at the time of his being in which, they see him in his true situation; at other times, and at other places, his rays falling obliquely, they are refracted, and therefore he is not seen in his proper place.

The same happens in regard to all the heavenly bodies, shining either by intrinsic or reflected light; for when that light falls obliquely upon our dense atmosphere, it is refracted; therefore we never see any celestial object in its true place but when it is in our zenith point.

To estimate the deviations of the rays of light, in order to ascertain the true place of a celestial object, is one of the most difficult parts of aftronomical calculation.

The more we consider the intricacies of the sublime sciences of Astronomy and Mathematics, the more do we admire the superior sagacity and perseverance of those who have gone extensively into those investigations; it requiring most astonishing strength of mind to bear such continual exertion of the mental powers, and a disposition the most tranquil to endure it.

It will be impossible in this part of my subject to avoid repeating occasonally what I have before advanced; for which unavoidable circumstance I hope to be excused.

You know that the refractive power of our atmosphere occasions that faint light, which in the morning we call Aurora, and in the evening twilight. Astronomers have estimated the degrees of refraction at every degree of altitude; and have found that they diminish in their degrees as the Sun approaches the zenith point. The latter part of this hypothesis is undeniable; but the former cannot surely be settled for all seasons at the same ratio, as it must depend on the state of the atmosphere.

In illustrating the effects of the refractive power of the atmosphere, I must refer you to *fig.* 3, *plate* 7, in which A B C D represent our Earth, E F G H its atmosphere, which decreases in density as it is more distant from the surface of the Earth, as has been discovered by the barometer; the mode of applying which to this observation I have previously communicated.

Suppose A, *fig.* 3, *plate* 7, to be the place of a spectator on our Earth, who is viewing the heavens; let S be supposed the Sun, which is not yet risen above the horizon, I K, of the spectator at A. When the Sun's beam, represented by the line from S to L, touches the atmosphere at M, it will be turned from its strait course, and be so refracted that, to the observer at A, the image of the Sun will appear at *s*, in the direction *s* A; by which means he will see that luminary before he has risen above his horizon I K. Again, when the Sun is above the horizon I K of the spectator at A, in the direction *s* N, the rays of light falling upon the atmosphere at N, and being refracted, are drawn towards the perpendicular A P, and appear to the observer at A in the direction Q A. When the sunbeam becomes a tangent to the atmosphere, as at R, then twilight ends; because the light does not fall upon our atmosphere. Refraction sensibly takes place when the Sun is about 18 degrees below our horizon. The circumstance of the Sun being about 18 degrees below our horizon, at the beginning of Aurora in the morning, and at the ending of twilight in the evening, has enabled mathematicians to calculate the sensible height of our atmosphere, but with certain limitations I imagine, allowing for the state of it, as it must sometimes be more strongly refractive than at others.

The state of the atmosphere is often very improperly expressed by those unacquainted with its nature, who, when their spirits are depressed, ascribe that effect to the heaviness of the air; whereas, on the contrary, it is produced by its rarity or lightness; which rendering its impressions weaker on our bodies, not being sufficient to counteract the spring of the air within them, the blood vessels and nerves become relaxed: but when the air is more dense, (as is always the case when we see the clouds and vapours buoyed up by it) its weight and elasticity balancing the internal air, braces our whole system, making us feel light and sprightly. The barometer evinces this very clearly, the quicksilver falling in it when the sky is obscured by clouds and vapour, and rising when these are dispersed, and the blue canopy of heaven encircles us.

As the state of the atmosphere must determine its power of refraction, the length of the time we have the image of the Sun refracted to our view, before he actually rises above our horizon, must be variable. When the sky is clear, by the greater density of the atmosphere, we see the Sun about ten minutes before he transits the horizon; but when it is cloudy, sometimes not above six minutes, the time differing according to the density; which difference, from the most dense to the most rare state, is found to be about four minutes.

We also see the image of the Moon before she transits our horizon, which in the state of the densest atmosphere is not more than two minutes; the orbit of the Moon making but a small angle with that of our Earth.

The duration of the Aurora and twilight depends on the obliquity of the ecliptic circle, or Sun's apparent path with the horizon, in the particular part where the Sun is, on account of the length of time he is passing through the 18 degrees below the horizon, which are counted perpendicular to it; and therefore the more oblique the direction of the Sun's path to the horizon is, the longer must it be in passing through that space. On the latter the duration chiefly depends, and is always calculated in proportion, allowing fomething, I should suppose, for the difference of refraction in reference to the different state of the atmosphere at different seasons.

Reflecting on the laws of refraction, that when the light falls from a rare on a denser medium, it is refracted towards the perpendicular of the denser medium, we are sensible that the object, by this law, must always appear higher than it really is; yet not actually so high as if refraction alone caused the deviation of the object from its true place; for parallax (the nature of which I shall explain to you) acting in a different direction, diminishes the height of the object.

It is not only refraction and parallax which cause our senses to be deceived in respect to the real situation of the heavenly bodies—the aberration or motion of light assists in the delusion, if not allowed for.

Let us first consider what relates to parallax, and the meaning of the term.

The arch of the heavens intercepted between the true and apparent place of a celestial object, (independant of refraction) is called its parallax.

The true place of a star is referred to the observation of it, as if taken from the centre of our Earth, which not being practicable, in order to ascertain its true place, the difference must be allowed for; as thus:—

Let A B C D, *fig.* 4, *plate* 7, represent our Earth, and T the centre of it. Suppose A the situation of a spectator on it, and E F G the orb or sphere of the fixed stars, also *efg* that of Jupiter, or any other planet. If the spectator, on the surface of the Earth at A, be supposed looking at the Sun, Moon, or a planet at *f*, it will appear at G, which is its apparent place; whereas that in which it would be seen from the centre of our Earth, at T, is the point F, which is its true place. The difference of place is called the parallax of altitude, and is the difference of the angles formed between two visual rays, and the perpendicular A E; one of the former being directed to the centre, and the other to the surface of the Earth.

The parallax of altitude is measured on the arch of the great circle, which is between the true place F, and the apparent place G. It is also called the diurnal parallax, as it depends on the diurnal motion of our Earth.

The parallaxes used by Astronomers are many; and which they are obliged to allow for, when taking the declination, right ascension and descension, longitude and latitude of the heavenly bodies; all of which differ in respect to the parallax. To define all of these by diagrams would be tedious, and is unnecessary to you, as reference to them is not requisite to any but those who study the abstruse mathematical demonstrations of these truths; who will not need a definition of them in this place;—therefore, having given you an idea of what is meant by parallax, I shall only define the meaning of its different applications in Astronomy.

Parallax of declination means the diminution of the declination of an object by the parallax of altitude. The parallax of altitude I have fully explained, to which you must transfer your idea, in respect to the meaning of parallax of declination, right ascension and descension, longitude and latitude.

Parallax of right ascension is reckoned on an arch of the equatorial, as you will perceive, when working the celestial problems annexed to these Lectures; the parallax of altitude increases the ascension, and diminishes the descension; these are both counted on the equatorial circle.

Parallax of longitude is an arch of the ecliptic, shewing how much the parallax of altitude increases or diminishes the longitude.

Parallax of latitude is the arch of a circle of latitude, whereby the parallax of altitude increases or diminishes the latitude.

There are also other parallaxes, one of which is the angle formed by two right lines, drawn from the Sun to the Earth, and from the Sun to the Moon, at either of their quadratures.

Another is called the parallax of the annual orbit of the Earth, which is the difference between the place of a planet, as viewed from the Earth and the Sun.

The parallax of altitude decreases the altitude of the object, as you may see by *fig. 4, plate* 7 ; in which F is the true place, being referred to the centre of the Earth at T, and G its apparent place, at which it appears to a spectator on the surface of the Earth at A, the true place being higher than the apparent ; which effect of the parallax of altitude being contrary to that of refraction, (which elevates the object's apparent situation) in some cases makes the effect of refraction very inconsiderable.

The parallax, as well as the refraction, being greater in proportion to the greater obliquity of the object to the horizon, they must both diminish in proportion as the object rises towards the zenith, where it is nothing ; therefore neither parallax nor refraction have any effect on objects in our zenith.

Parallax does not affect our observations of the fixed stars, as viewed from our Earth, on account of their immense distance from it ; the Earth being but like a point, its diameter is as nothing, as well as that of the Earth's annual orbit ; from which cause it is that Astronomers cannot estimate the sizes or distances of the fixed stars, not being able to form an angle with any of them, either by the semi-diameter of our Earth, or by that of its orbit.

A parallactic angle, or parallax, which means the same, is the angle formed in the centre of the Sun or a planet, by two right lines drawn, the one from the centre of the Earth, and the other from its surface.

The parallaxes chiefly used by Astronomers, are those which arise from viewing the object from the centre of the Earth ; and considering it as viewed from the centre of the Sun, and also from the surfaces of the Earth and Sun, and from all compounded ; these are used to discover the sizes and distances of the planets of our system, both in respect to the Sun and our Earth. The difference between the place of a planet, as seen from the Earth and Sun, is called the parallax of the Earth's annual orbit. The nearer a planet is to the Earth, the greater is its parallactic angle.

Suppose, in *fig.* 5, *plate* 7, T to be the centre of our Earth, and A B C its surface, D E F G part of the orbit of Jupiter, and H I K the same portion of the orbit of Mars.

The angle B E T, made by the more distant planet with our Earth, is less than B I T, that made with the nearer, as may be seen by the portions of the semi-circle A B C between each.

In regard to the parallax of the Moon, several things are to be allowed for, her meridian altitude, the equation of time, her longitude, latitude, declination, &c.

The Sun, although an immense body, yet being so far removed from our Earth, has but a small parallax with it; and therefore Astronomers have recourse to the parallaxes of the planets nearest to our Earth, Venus and Mars, from which that of the Sun is deducible, as by knowing the motions of the Earth and planets, they are able to ascertain the distance between them and our Earth, for any part of their orbits, and also between them and the Sun; by which they are enabled to find that of the planet from the Sun, at the time of its conjunction with, or opposition to, that luminary; in the former case, by adding the distance of the planet from our Earth at that time, to its distance from the Sun in that part of its orbit, the sum shewing the distance of the Sun from the Earth, when the Earth is in that part of its orbit; also by observing the apparent diameter of the Sun at that time, and knowing the diminution of objects according to their distances from the place of observation, they discover the size of the Sun; as they do that of all the planets by the foregoing observations.

When Mars is in opposition to the Sun, knowing that he is twice as near to our Earth as the Sun is, by observing his parallax with our Earth, which must be twice as much as the Sun's, they discover the parallax of the Sun, and are able to make the same inferences as in the former case.

Cassini, by observation of Venus, when in her inferior conjunction, (at which time he knew how much nearer she was to our Earth than the Sun, and consequently that her parallax must be so much larger with our Earth than with the Sun) he found the Sun's parallax with our Earth was 10 seconds, and that his mean distance from it was equal to twenty-two thousand semi-diameters of the Earth, or eighty-one million of miles; the semi-diameter of the Earth at the equator being computed at three thousand nine hundred and seventy-eight miles.

The distance of Venus from the Earth, at the time of her being in a direct line between it and the Sun, is less than her distance from the Sun; and her distance from the centre of the Sun is found by her passage over the Sun's disk, which is called the transit of Venus.

The transit of Venus has excited the particular attention of Astronomers. That of Mercury at the time of his inferior conjunction might also be observed, as at that time he appears like a dark spot on the Sun's disk; but on account of his nearness to that luminary, the difference of the parallax between him and our Earth, and the Sun and the Earth, is so little, as not to answer the purpose, not being sufficient to make any deduction from. But the parallax of Venus being nearly four times as great as the solar parallax, on account of her being so much nearer to the Earth than Mercury, and further from the Sun, it causes great difference in the time in which she is seen to be passing over the Sun's disk from different parts of the Earth.

It was Dr. Halley who was so particularly anxious to engage the attention of Astronomers to this mode of determining the Sun's parallax, and for which purpose he furnished the Royal Society, in the year 1691, with an account of the particular periods when such

Q

transits would happen, assuming the position, that if the interval in time, between two central transits of Venus over the Sun (Venus does not always pass over the centre of the Sun's disk in every passage, on account of the obliquity of her orbit to the central solar ray,) could be calculated to a second in two places of different degrees of longitude, the Sun's parallax might be calculated pretty accurately, allowing for the difference of time at which it should appear at each, on account of the diurnal motion of the Earth. As thus.—Suppose one observer is situated 90 degrees west of the other, it will be six o'clock in the morning at the former situation, when it is noon at the latter; in agreement with which local circumstances, that allowance must be made. Then let us suppose that Venus transits the centre of the Sun at the former place at 7 minutes past 6 o'clock, and at the latter at 12 o'clock at noon. Deduct 6 hours for the difference of longitude, and the remainder will be 7 minutes of time; which time being converted into parts of a degree of the equator, is called Venus's horizontal parallax with the Sun. I shall illustrate this part of the subject of Astronomy in the manner of Ferguson, who has familiarly explained the mode by which those grand objects of the size and distance of the Sun have been ascertained, and which will give you a good general idea of parallax. It is called, taking the horizontal parallaxes of the Sun and Venus, because it refers only to those bodies and the Earth in a horizontal position, and is the mode by which all the parallaxes of the heavenly bodies of our system are worked, in order to discover their real distances from each other; which being thus known, their sizes are estimated by their apparent diameters at those distances.

The Sun, on account of his immense distance from our Earth, when viewed from two places on its surface, will appear nearly in the same point of the heavens; but Venus, being less than the Sun, and almost four times nearer to the Earth, when she passes over the Sun's disk, will appear at the same time to be on different parts of his surface, to two spectators placed on different parts of the Earth.

In *fig.* 1, *plate* 8, S is intended to represent the Sun, V Venus, and A B D E the Earth. Suppose one observer to be at A, another at B, and a third at D, viewing Venus at the same instant. To the observer at A, Venus will appear upon the Sun at F, as she will be seen by him in the right line A V F. To the observer at B, she will appear upon the Sun at G, in the line B V G : and to the observer at D, she will appear on the Sun at H, in the direction D V H. These positions are very clear, as the lines represent the visual rays of the spectators, observing the phœnomenon from different parts of the Earth's surface.

The elliptical form of the planets orbits causing them to be at different distances from the Sun, and from each other, in different parts of them ; to know their mean distance from the Sun, it becomes necessary to compare their least and greatest distances, and take a mean between them ; in order to which, observation must be taken of the transit of Venus at two different distances from the Sun, as well as when the Earth is in different parts of her orbit ; and when the mean distance of one planet is known, all the rest may be discovered by that standard.

Ferguson has shewn what the difference of the parallaxes is, in two different parts of the orbits of Venus and our Earth ; we will, therefore, pursue his explanations, referring to the diagrams by which he illustrated them. *Fig.* 2, *plate* 8, has the Earth, *a b d e,* placed nearer to the Sun, S, than in *fig.* 1, *plate* 8, consequently Venus is nearer to the Earth ; and the arc, *a d,* intercepted between the two observers placed on the Earth's surface at *a* and *d,* bears a greater proportion to the distance of Venus from the Earth, than it does in *fig.* 1, *plate* 8 ; i. e. the angle formed between them is larger, because their distance from each other is less.

As the space, *f g h*, in *fig*. 2, over which Venus passes, is longer than F G H, *fig*. 1, the planet will be longer in passing over the former than over the latter; for the nearer the Earth is to the Sun, the greater will be the space through which Venus will appear to move on the Sun's disk; which is occasioned by the real motion of the Earth on its axis, on which the observer is placed, and of course, by which motion he is carried along: therefore, you perceive, that the nearer the Earth is to the Sun, the greater will be the space that Venus will appear to us to describe on its surface, and consequently the longer will she be in passing through it. The real motion of Venus in her orbit, compared with the time she takes in passing over the Sun's disk, gives her parallax, as I have previously informed you.

These positions we are able to authenticate by a quadrant to our intire satisfaction.—Having explained the principles on which the sizes and distances of the bodies of our system are known, I shall next treat of the mode of computing their motions in their orbits.

By the laws of motion it is discovered, that the velocity of a body moving round a centre, will be in proportion to the squares of its distance from its centre of motion, compared with its size; therefore, the real size of the planets of our system being known by their apparent size at known distances, and those distances being also ascertained, their velocities have also been calculated, and which have answered accurately to their known periods.

Let us consider the effect of the apparent diameter of objects, at different distances, by recurring to those of the Sun and Moon. These appear nearly of the same magnitude as viewed from our Earth, although their real sizes are very different; this apparent equality of size is occasioned by their different distances from the Earth. The Moon being only two hundred and forty thousand miles from us, whereas the Sun is ninety-five million of miles removed from our Earth; and the bulk of the latter exceeding that of the former sixty-four million times, makes them at that differ-ence of distance both appear of the same size; the apparent dia-meters of objects diminishing exactly as their distances increase.

Although we have not entered along with those great men (who have applied these known laws to the grand objects which caused the investigation) into all their complicated and tedious processes, yet we are able to form a sufficient idea of their mode of calcula-tion, to give our assent to the practicability of ascertaining the size, distances, and motions of those heavenly bodies which are not too remote for their scrutiny. Their indefatigable zeal and accuracy in prosecuting their speculations is evident, by the correctness with which they calculate and foretell the appearances of celestial phœ-nomina for different periods.

Those who wish to amuse themselves with applying the previous intimations of Astronomers, in order to satisfy themselves of their authenticity, and to examine into the wonders of the celestial re-gions, I recommend the use of White's Ephemeris, or the Nautical Almanac, which contain every thing necessary for observation; and the time being equated for every day in the year, makes it par-ticularly useful. I purpose annexing to these Lectures a Key to the Ephemeris, in order that its use and application may be pro-perly understood by those not pre-acquainted with the information it is intended to convey.

Having explained two of the causes of our not seeing celestial objects in their true place, excepting when they are in our zenith —parallax and refraction—the remaining one must not pass quite unnoticed, although not very essential to us, being but trifling in its effects, and therefore scarcely sensible.

This, which is called the aberration of light, is occasioned by the motion of that fluid compared with that of our Earth; and is referred to the fixed stars more particularly. It was discovered by that great Astronomer Dr. Bradley, in the year 1725.

That there was an apparent difference in the places of the fixed stars, when observed at the same time for many nights together, had been previously noticed by Dr. Hook;—but the cause which produced this change was not discovered till Dr. Bradley unveiled the myftery.—It was long before this gentleman could ascertain the fact he wished to discover, as it required instruments and time-pieces of peculiar accuracy to make the observations necessary for determining it. The particulars of this subject are related nearly in the following manner by Dr. Hutton in his Dictionary:

Dr. Bradley was accompanied and aided in his researches on this subject by Mr. Molineux. The instruments they used were con-structed by Mr. Graham, on the accuracy of which was their chief dependence.

On the 3d of December, 1725, they remarked the bright star on the head of Draco, as it passed near their zenith, and carefully took its situation by the instrument, noting the time of observa-tion; they repeated this process on the 5th, 11th, and 12th days of that month, at the same hour, when there appeared no essen-tial difference in its situation.

On the 17th of December, amusing themselves with the instru-ments, they again viewed this star, not expecting to see any dif-ference in its place—but were much surprized to find that it passed a little more southerly than before.

They knew that this change of place could not proceed from the nutation of the Earth's axis, (which motion I shall have a better opportunity of explaining to you than the present offers) therefore they suspected the accuracy of their instruments; but on the 20th of December they repeated their observations of this star, and perceived that it was removed still more southerly. This dis-sipated their former fears, and animated them to a further scrutiny Perceiving that the departure of this star from the pole was gra-dual, they became very anxious to discover by what regular cause this was produced; for which purpose they observed how far it was removed at each observation, till it arrived at its most southern limit, which it reached in March 1726, and where it was removed from the place of their first observation of it, Dec. 3d, 20 seconds more southerly.

In this situation it remained till the middle of April, when they first perceived its return towards the north. The beginning of June it had arrived at the same point in which they noticed it in December; and from that time it pursued its northern direction till September; when, on its becoming stationary, they perceived that it had advanced 20 seconds nearer to the pole than it was in December and March.

From September it returned towards the south, till it had regained the same situation that it was in on the day they first made observation of it, allowing something for its difference of declination on account of the precession of the equinoxes.

Without following the learned Doctor through all his observations, I shall just give you the result of them.

After comparing all their observations, with various conjectures respecting the cause of those regular deviations they had perceived; at last it occurred to Dr. Bradley, that it might proceed from the progressive motion of light, referred to the Earth's annual motion in its orbit. If, says he, light be propagated in time, the apparent place of a fixed object would not be the same when a person is moving in any other direction than that of the line passing through the eye, as it would be if the eye were also at rest, but would appear in different directions, according to that motion of the spectator.

This is too evident a position to be doubted; and therefore we have now two observations by which we are assured of the velocity of light, and that it is progressive like that of all other bodies in motion.

Thus having found that the apparent places of the fixed stars, by the velocity of light, differed about 20 seconds from their true places, **Dr. Bradley**, by trigonometry and calculation, proved that the velocity of light is about ten thousand times greater than the velocity of the Earth in its orbit; for knowing the mean velocity of the Earth's annual motion, and comparing it with the apparent changes in the places of a fixed star, he could compute what were the relative velocities of light and the Earth in its orbit; as the difference between the apparent places of the stars would be in proportion to the two velocities.

These inferences making the velocity of light to be the same as it was before estimated by observation of the eclipses of Jupiter's satellites—who will venture to confute the validity of as clear a fact as ever was demonstrated?

Having conducted you thus far, I shall close my Lecture, that your imaginations, not being too crouded, may have room to exercise themselves, by reflecting with due attention on the wonderful and benevolent laws which govern all those processes of nature, with which you are already acquainted, before I introduce you to the other wonders and beauties of the creation; so preparing you by degrees to contemplate such a glorious assemblage of divine things as you will have exhibited to your understanding in the course of these Lectures.

By what I have already delivered, you must perceive the advantages of investigation, and of properly examining into the effects of the minutest subject; by which you have heard what great and useful truths have been discovered.

These exercises of the mind produce reflections the most enlivening;—hence also the judgment is strengthened, and ample sources of amusement and instruction procured, and by which you will avoid the languor attendant on an inattentive observer of nature, who may be said to vegetate rather than live; for what is life without knowledge, or understanding the things by which we are surrounded, and from which such inferences are deducible, as alone can elevate the mind, and make us feel the important rank we hold in creation, with the certainty of a future state of more glorious existence.

If any one deny a vacuum, let him look into the mind of an unreflecting person, and he will find one; not indeed exactly what he may be looking for, but figuratively so; and to pursue the metaphor, the stone which has been so long sought for, may be found in the bosom of the natural philosopher, whose mind, constantly exercised on subjects the most interesting by the ideas they convey of divine things, feels a tranquillity unknown to the uncultivated in the sciences.

The mind has a constitutional activity, which cannot be totally suppressed; but sometimes it is exerted on subjects which disgrace its inherent capacity, such as examining into the affairs of our neighbours, for no other purpose but to gratify an impertinent curiosity; such activity bespeaks a mean understanding, to say no worse. I do not mean that the good or bad actions of our neighbours should be indifferent to us, or that the one should not receive our commendation, and the other our censure; as such passiveness may subject us, in the former case, to the imputation of indiscrimination, and in the latter imply our sanction. Were these discriminations more generally acknowledged, I believe the world would be much better than we find it; for whilst riches shield a vicious person from public contempt, I fear virtue will not intirely triumph—except in its own consciousness.

Do then, my dear pupils, shew your superiority, independency, and purity of mind, by shunning the vicious, however exalted or aggrandized, and by countenancing virtue under all circumstances; but do not mistake the meaning I wish to convey by the word independency, and imagine that those whose superior rank renders them deserving of particular respect, by including the advantages resulting from a polished, rational, and religious education, are not intitled to it; that is not to be implied: as to distinguish such people by particular attention, and to cultivate their acquaintance and friendship, is but a just tribute to their merits, and our judgment; whereas the withholding them, and affecting to disregard their opinion, implies gross ignorance, and want of cultivation in ourselves; for sure every person of sense and good breeding will, with pleasure, subscribe to the forms of society, to procure the advantages resulting from the friendship and conversation of a polished and well-informed person, which they must so greatly prefer to those of a rude and illiterate one.

Besides the advantages just mentioned, you will also in general find more true religion and virtue in the higher walks of life than in the lower, as vice is the growth of a foul and neglected soil, and virtue of the purest and most highly cultivated; for, look into the minds of those who are vicious, either in high or low life, and you will find them equally gross and unenlightened. Therefore, do not confound the means with the end, but only regard the means when the end is obtained.

As you all must, I am sure, admire virtue in others, you will, I hope, practise it yourselves,—and as well for its own loveliness, as for the advantages it affords.

> And if you pant for glory, build your fame
> On that foundation, which the secret shock
> Defies of envy and all-sapping time.
> The gaudy gloss of fortune only strikes
> The vulgar eye; the suffrage of the wise,
> The praise that's worth ambition, is attain'd
> By sense alone, and dignity of mind.
> Virtue and sense are one; and, trust me, he
> Who has not virtue, is not truly wise.   ARMSTRONG.

END OF THE FIFTH LECTURE.

LECTURE VI.

OF THE FIXED STARS; AND A VIEW OF THAT SUBLIME FIELD
OF INVESTIGATION, THE UNIVERSE.——THE NORTHERN
LIGHT, OR AURORA BOREALIS.——CHANGES IN THE
FIXED STARS.——OF MOTION AND GRAVITY.

FROM Astronomy we learn the immensity of that Being, who
could thus perfect his work in wisdom,—and control its various
operations by his power.

Our wonder and admiration is naturally extended beyond all
bounds by the sublime objects of our present consideration; yet if
we examine through all nature, from the minutest object to the
most enlarged, we shall find equal cause for astonishment.

In the survey of our system, we perceive the order and influences
even of those we cannot examine into; and reasoning analogically,
we acknowledge the idea of worlds on worlds innumerable, dispersed through the vast extended universe;—ideas too large for
finite conception, which must be governed by finite laws,—laws
which are totally inadequate to the real aggregated conception of
this conflux of systems on systems impelled by divine command,
and immerged in the bosom of immensity.

To comprehend the real excellency of the design, we must examine its parts distinctly, by which alone we can form a right notion of it: as thus—When we consider all those bright luminaries,
the fixed stars, each as belonging to a particular system;—that it
has worlds circulating round it, as that of our system has; and that
they are kept in their proper stations, receiving a due proportion

of the influences of this great central body—that each system is kept from infringing on its neighbouring system; so that they suffer no impediment from each other in their various independent affections; and lastly, that these worlds are probably peopled with creatures whose necessities must be amply provided for, if they exist at all, by that great and good Providence, who has created all things—how does our admiration of the whole design increase! Even by diminishing the object, we enlarge its importance: for as the investigation of the organs of the minutest insect but serves to increase our astonishment by the skill displayed in them—so does the usefulness we discover in each part of the universe, when separately considered, raise our ideas of that great and good Being, who, in exhibiting the sublimest spectacle of his power, has not neglected the minutest want of all his creatures.

Smart has beautifully pourtrayed the sufficiency of the smallest creatures to supply the exigencies of their existence, not only by the construction of their organs, but also by the instinctive foreknowledge they are endowed with; which he has exemplified in the little ant.

" Conscious that December's on the march,
Pointing with icy hand to want and woe,
She waits his due approach, and undismay'd
Receives him as a welcome guest, prepar'd
Against the churlish Winter's fiercest blow.
For when as yet the favorable Sun
Gives to the genial Earth th' enliv'ning ray,
All her subterraneous avenues
And storm-proof cells, with management most meet
And unexampled housewifery, she forms:
Then to the field she hies, and on her back,
Burthen immense! she bears the cumb'rous corn.
Nor rests she here her providence, but nips
With subtile tooth the grain, lest from her garner
In mischievous fertility it steal,
And back to day-light vegetate its way."   SMART.

Plate VIII.

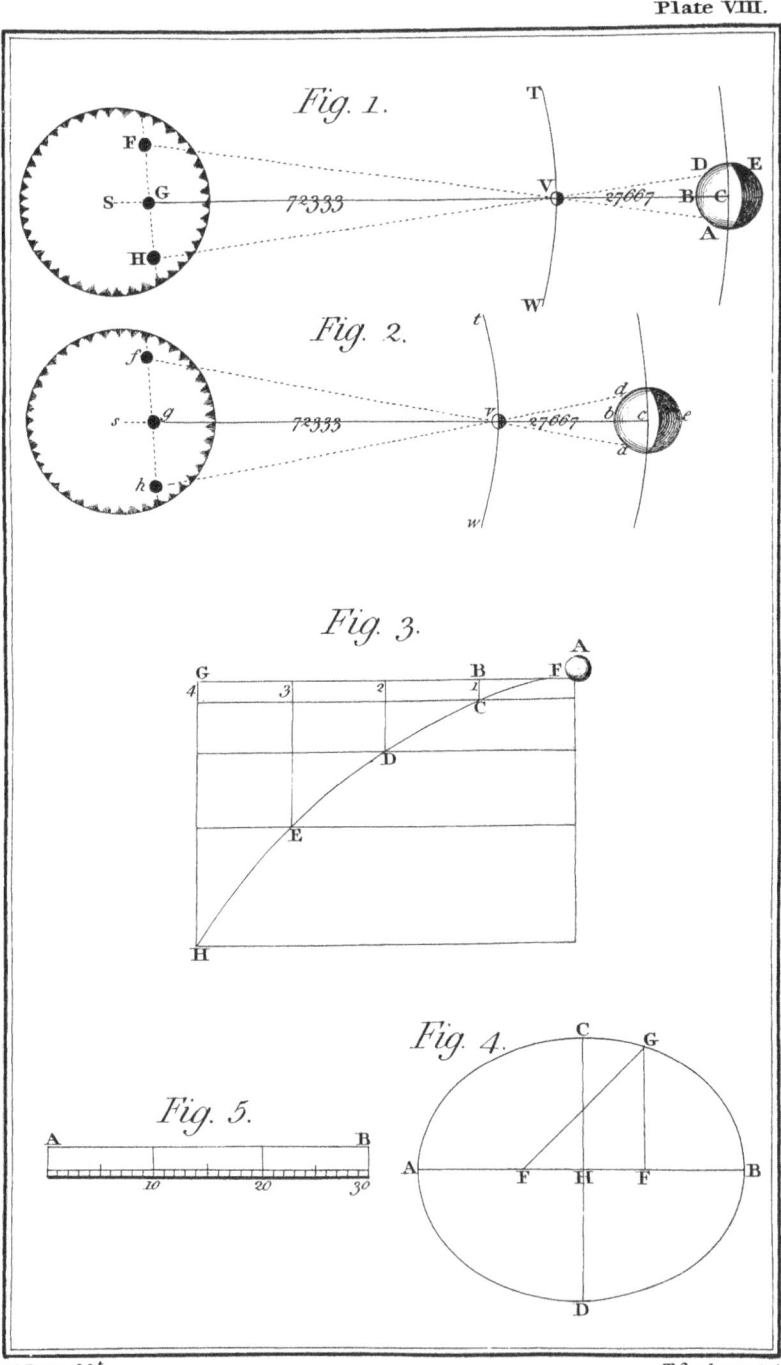

Fig. 1.

Fig. 2.

Fig. 3.

Fig. 4.

Fig. 5.

Thus we see that the attributes of the Almighty are as evident in the smallest, as in the largest portion of his works; and like the particles of light, the more they are separately considered, the brighter do their native beauties appear.

The sublime imagery of the universe, usually presented to us by writers on this subject, although extremely beautiful and noble, becomes more valuable, and creates more love and admiration of the Creator, when analized, than when considered only in the mass —without reference to its divisible excellence.

Dr. Herschel, a gentleman of great abilities, which he employs most gloriously, and for which application of them he deserves more praise than I can bestow, has furnished us with many curious observations, which may with propriety be admitted, as his indefatigable exertions, and astonishing instruments, aided by the most penetrating discernment, validate all his assertions, and render his inferences worthy attention.

This great man, like all superior geniuses, advances his opinions with great caution, and never asserts any thing positively but what he is assured of.

He conjectures that the numerous hosts of systems dispersed throughout the wide extended firmament, are connected together by the mutual contact of their circumambient atmospheres, altogether forming one great sphere; in the manner I suppose he means, as we observe of a large drop of water, which is composed of a number of smaller globules, kept together by the mutual cohesion of their external parts—occasioned by their centripetal forces, which are equal to each other in each, and exerted in every possible direction; so that by counterbalancing each other, an equilibrium prevails, which assimilates them, and connects them together, forming one perfect sphere of immense magnitude. Knowing that the fixed stars shine by their own light, and not by reflecting the light of our Sun, we with reason conclude that they are each of them of the same nature with our central body; which being admitted, we naturally infer that they give light and life to other systems of planets, as that of our Sun does to its circumfluent worlds.

From the known divergence of light from a centre, it is that we know that the fixed stars cannot shine by reflecting the light of our Sun, as at their great distance from it, the Sun's rays cannot reach them, being all dissipated long before they could arrive at those remote regions, and therefore none of their effects can be experienced by those bodies. Again, we are sure that the fixed stars are of no use to our Earth, or the other planets of our system, and therefore with propriety we conclude they are intended to answer the same beneficial purposes with our Sun to other worlds. As what is there in nature which does not evince usefulness as well as beauty?—Nothing. Then would it not be contrary to common sense, and common observation, could we weakly suppose that God has created such immense bodies without that great design, which we perceive evinced in all he does—of connecting beauty, harmony, and the most extensive usefulness, with the most unbounded benevolence; so that we cannot say which of these is the most prominent characteristic of his works. This must convince every sensible mind of the real nature and appointment of the innumerable hosts of fixed stars; and raise in us the most profound love, gratitude, and adoration of a Being, so wise, so good,—possessed of power so unlimited, of glory so resplendent, of benevolence so universal—and which are never ceasing, never ending. I am so animated by the subject, that I could go on for ever, enumerating those impressions of an all-wise and bountiful Creator. I know not how the Almighty performs all the wonders I behold, nor do I vainly attempt to dive into what is, I know, far above human comprehension, never being intended for the introspection of earthly existences. A time will come, when we shall behold the works of God as they really are, resplendent in truth and glory. In the mean time we must endeavour, by purifying our minds, to fit them for the participation of the joys which such divine contemplations and participations afford those who are partakers of them.

Had the Almighty blessed us with every thing fitting for our earthly existence only, we should have been certain that this state was to terminate it; but he has not been less provident in supplying us with all that is requisite for a future state of exaltation; and in order to our attaining this perfection, he has furnished us with a strong sense of right and wrong, even so strong, that those who act the most basely, acknowledge it, although they act in contradiction to its intimations. Why should we have this instinctive knowledge infused into us, if we were not intended for a better state than this? or why were we created at all? and, above all, why furnished with a soul, that active principle, whose energies are never ceasing, never suspended?—These intimations can never be mistaken.

The brute creation is not sensible of good and evil, excepting as far as respects their corporeal necessities; therefore we conclude that they are not designed for a future state of reward or punishment, but were created for our use, though not for our abuse; which latter is a perversion too common, and on no account justifiable, and for which, like all abuses of the dispensations of providence, no doubt we shall one day be called to account.

Some people doubt of a reward of good and evil deeds, because they wish there may be none. If there were to be no reward, wherefore should we be placed on this stage of existence at all? We might at once have entered on our perfect state, had we been fitted for it; and why, if all our deeds are equally good, were we furnished with an inward sense of right and wrong? Admitting the necessity for such a monitor, we also must admit that there is a cause for it; and that the Almighty, who has so fully evinced his superintending and preventing goodness in the natural world, would not leave us destitute of what concerned our spiritual well-being.

The Almighty has perfected us in all that is necessary to our obtaining everlasting happiness, by having sent us his only Son, to redeem us from death; and also to teach us how we ought to act in this life, in order to attain to a better. No means could so fully have confirmed us in the belief of a future state, or in the practice of virtue, as the appearance and divine character of our Saviour. He was incumbered with a fleshly body, yet kept it in subjection; no malignant passions had existence in his mind, yet he suffered the most cruel tortures of both body and mind; and all this to shew us how we ought to conduct ourselves. He not only took upon him our flesh, but patiently endured the pains of mortality, the greatest bodily sufferings. Let us then endeavour to copy this bright original; being assured, that by conducting ourselves according to his example, we can alone be happy in this life, and obtain the joys of futurity. I recommend those who doubt, to peruse Payler's Truths of Christianity, in which they will find ample sources of consolation and conviction. I will not invade the clerical province, by offering to establish truths, and confute erroneous opinions of religion, in this place, but conclude this part of the subject, by exhorting to guard against being misled by common opinions, by examining into the Truths of Christianity; and this I think it my duty to caution my pupils against, being conscious that any innovation in religion, like that in a state, is but the forerunner of its total destruction. As when we begin to forsake God, he will forsake us; therefore we should guard against schism, not only for its own sake, but also because it too frequently leads to, and terminates in, the most deplorable of all evils, skepticism.

But to return to the contemplation of those objects, from which such delightful evidences are deduced, and of which we can only judge by analogy. That effect of the light of the fixed stars which is called twinkling, proceeds from the vibration in the particles of light, produced by the particles of the atmosphere which interpose between us and them, and is peculiar to intrinsic light; as when light is reflected in the mass as it were, its parts are not so divided from each other, and therefore the atmosphere does not communicate to them that undulating motion. The stars appearing colourific when viewed through certain telescopes, is owing to the light being refracted by the glasses; and as all the particles of light are not equally refrangible, some of them will not unite at the focus of the glass, but remain in their separated state. The quantity of light not being augmented by glasses, as the surface of the body is, may occasion the light of the stars to appear duller through them, than when viewed by the naked eye; the refraction I have mentioned, may also contribute to the diminution of lustre, wanting something of that due mixture which constitutes a bright light. The former is a long established position, the latter my conjecture; of the propriety of which, those acquainted with the nature and properties of light, and the effect of glasses, can best judge.

The fixed stars, as the name indicates, are perfectly at rest in respect to each other, as may be known by observing the relative position between any two stars in respect to any part of the Heavens; as they always appear in the same situation, wherever observed from.

The reason of our supposing the stars to be set in a concave sphere, arises from the indefinite extension of the universe; and our viewing them from a sphere of light, as it were, which is formed within our eye. The fixed stars are not supposed to be equidistant from us, but to be placed at different degrees of remoteness, which occasions some of them to appear larger, as viewed from our Earth, than others; those which are nearest, of course, appearing largest; and those most remote, the most minute; therefore, although we apply the terms expressive of different degrees of magnitude to

S

them, yet they refer to appearances only, as they may be of the same magnitude for ought we can know to the contrary

Dr. Herschel, by means of glasses of immense power, has discovered that the galaxy, or milky way, as it is called, and which runs irregularly and obliquely to the zodiac, and crosses it in Cancer and Capricorn, always keeping the same situation, passing through Cassiopeia, Cygnus, Aquila, Perseus, Andromeda, Ophiucus, and Gemini northerly, and by Scorpio, Sagittarius, Centaurus, the Argaunatis and the Ara, southerly — consists of an innumerable quantity of fixed stars. This gentleman conceives our Sun to be one of this extensive stratum of luminaries. These Suns of the universe may, or may not, all be of the same size, as the Almighty could proportion the planets to an animating body of any dimension.

The Doctor says, that although the Sun, which regulates our system, is in the centre of it, in respect of us, yet that we should not consider him as in the centre of the universe, as most probably our system is composed of not one hundred thousandth part of the whole creation; for in the most crouded parts of the milky way, he has seen 588 stars pass through the field of his large telescope in the space of one minute.

The Aurora Borealis is a meteor only seen occasionally, but then mostly in the northern part of our horizon, in the northern hemisphere. It appears chiefly in spring and autumn.

The opinions of Astronomers concerning it are various, and differ so widely, that it is difficult to conjecture which is right, as they may all be wrong; yet as I think one has the preference, on account of its being deducible from known principles, I fhall venture to advance those circumstances which appear to validate it; but at the same time will not attempt to decide for others, and therefore I will give them the opportunity of making their own inferences of some of the most feasible ideas entertained of this phœ-nomenon, circumstantially relating the varieties observed in its appearance.

The Aurora Borealis has by some been ascribed to a subtile effluvia, which entering at one pole is ejected at the other with violence, in order to restore an equilibrium.

By others, to the effects of the Sun's atmosphere. Various other hypotheses have been formed, but nothing certain is ever likely to be understood of this extraordinary and partial illumination.—We may amuse ourselves as much as we please on this subject, but we must be satisfied with conjecture after the greatest stretch of our sagacity.

This light sometimes takes a horizontal direction, like the break of day; at others, like fine luminous beams, well defined, and of dense light, accompanied by a quick motion, which is always succeeded by sudden flashes pointing upwards in the same direction with the beams; this, in my opinion, gives some degree of sanction to the idea of their being the electric fluid passing off at the north pole, particularly as other circumstances tend to identify them—such as these appearances happening indeterminately, which bespeaks accumulation; for although particular circumstances may cause them to happen in some respects periodically, yet not entirely so, the transmission not happening till a certain quantity is accumulated—and when redundant, its being affected in the same way as the charge of an electric jar is when too replete—by bursting from the substance which contains it, thereby restoring its equilibrum.

All the testimonies we bring to authenticate this opinion, invalidate that theory which ascribes these appearances to the atmosphere of the Sun; for if they proceeded from that, they would be more constantly seen, and not liable to such irregularities in their form, and in their manner of acting.

The flashes which succeed the beams of the Aurora Borealis emit a fainter light, and are much broader, than the beams themselves; these flashings often continue for hours with intermission, but gradually grow weaker and weaker till they totally disappear. Their flashing only at intervals seems to imply their exchanging powers with each new cloud, which the revolution of our Earth on its axis brings them in contact with; all these effects are similar with those that would actually ensue from the electric fluid, yet it is impossible to assert positively that it is the same.

Sometimes the Aurora appears like arches, nearly in the form of a rainbow, reaching from one point of the horizon to another. The arches always cross the magnetic meridian at right angles, tending to the east and west points of the compass.

Those beams which are perpendicular to the horizon are so in respect to the magnetical meridian, and therefore we may suppose that this meteor has some affinity with the magnetic fluid; that the electric fluid, and that which causes magnetism, have affinity, we know by experience,—passing the electric charge of a battery, i. e. (of a number of charged jars connected together) through a magnet, inverts the poles—and polarity may be given to a piece of steel by electricity.

The circular direction taken by the beams at times has not been accounted for, but most likely it depends on the known laws of motion.

The light of this meteor being differently condensed in its appearance, at different times, may be occasioned by the different state of that part of the atmosphere through which it passes, and that by which its light is transmitted to us.

The sudden disappearance of certain stars, and the appearance of new ones, has excited the attention, and exercised the genius, of some of the most celebrated Astronomers both ancient and modern; as has also the apparent periodical changes of brightness in some of them.

Most of the theories advanced of these phœnomena are deduced from considering the stars, or Suns of the universe, of the same nature with ours, and made up of heterogeneous particles, which are capable of cumbustion at the surface of the body they compose. —Some have supposed the absence of these luminaries to be occasioned by the inflammable matter being decomposed, by which the ignition of course ceasing, there can no longer be any luminous appearance.—And also, that as combustion must continue some time before ignition takes place, this causes the appearance of new stars.

The former part of this theory I think defective, and the disappearance of certain stars more naturally accounted for by those who imagine their absence to be occasioned by a planet of the system that star illumines eclipsing it in its revolution, at a particular time of the nodes, and thereby hiding them from our view— than by those who attribute this obscuration to a default in the constitutional allotment of the essential properties which cause combustion.—I give you the former opinion as well as the latter, because it is equally popular; but still I think it improbable, not being consistent with the never-ceasing influence of the Sun, which illumines our system; and were this circumstance of defection to take place, what would become of that system which that Sun used to animate?—it must be totally annihilated. I do not say that this is impossible, by the will of God it might so happen; but I cannot with satisfaction subscribe to its being a natural and inevitable consequence, or by any means deducible from what we observe of the combustion of substances on our Earth, to which it may have no analogy; as I think we cannot with propriety explain the operations of infinite power upon principles deduced from finite circumstances.

The brightness of the fixed stars being sometimes greater than at others, most probably is occasioned by the spots on their surfaces, the same as are observed on the face of our Sun; and these also may occasion the appearance of what we call new stars; for when the surface of one of these Suns is much obscured by the spots on its disk, the quantity of light may be so much diminished, that on account of the immense distance of that star from our Earth, we may not see it for a considerable time.

Before entering into a nearer introspection of our system, it will be necessary to give you some further information of the laws of motion and gravity; without which, you would not be able to give your assent to what I am about to treat of; but when informed of the laws of motion, you will freely and instinctively subscribe to the facts presented to you of the motions of the planets round the Sun; and other motions referable to the Sun, the planets, and the satellites. And by the observation of the wise allotment of powers to each included in this instruction, you will be doubly gratified; as by comprehending the laws by which the great machine is actuated, you will discover the nice adjustment which nothing but infinite wisdom could ordain; and thus, by being able to estimate the effects, instead of diminishing the importance of the plan, you will deduce from it the most consolatory and elevating ideas, of the influence, the glory, and the goodness of God; who, if he were to with-hold his continual supply of animating matter, could in an instant check the planets in their courses, and produce universal disorder in a system, the order and influences of which are now so harmoniously beautiful.

The Almighty has been pleased to appoint motion his chief agent in all his operations, as is evident by all the phœnomena of nature, which universally obey the laws of motion, and are perfected by them.

It becomes therefore essential to those, who wish to comprehend the changes and operations perceived in that part of the great system of the universe, exhibited to them, to understand the known laws of motion; and the importance of this knowledge is evident in the advantages which have been procured, by attending to those laws, as well as by reference to the deficiency in the ancient Astronomy, from the universal ignorance of those laws which then prevailed.

The laws of motion are called laws of nature, because they are rules observed by all natural bodies.

Sir Isaac Newton has defined the laws of motion, and confined them to three states of dependency. His first law is, That every body perseveres in the same state, either of rest (or if impelled forward by one force only) in an uniform rectilineal motion; unless in the first case it be compelled to move by some power acting on it, or in the latter case, by a force in a contrary direction from the projectile, which it has received.

Thus we perceive that a body receiving a certain impulse, in a line strait forward, would pursue that course for ever, were it not for some other power retarding that motion.

In regard to substances on our Earth, they are retarded by two powers, one the resistance they meet with from the air, and the other gravity. The first opposing the progress of a body which has been impelled forward, and the other drawing it from its straight course downwards, by which motion its course is altered. But in the ethereal regions, where there is no sensible resistance, the planets perpetually revolve in their orbits, only the direction of their motion is continually changing from the force of gravity or attraction, which, in one part of their orbits, accelerates their motion, and in the other retards it.

Sir Isaac Newton's second law is, that the motion or change of motion, in a moving body, is always in proportion to the force impressed on it, to cause it to change its motion, and in the same direction of the received impulse. Thus, we understand that a double force impressed on a body, to cause it to move, will make it move with a double velocity, and so on.

All change of motion will retard the direct motion of a body in proportion to the force exerted to turn it from its straight course; i. e. according to the difference of the force exerted in the two directions. If I impel a body forward into the air with a certain force, the power of gravity will be continually drawing it from its straight course; and as the power of gravity in this case, is to my strength in projecting the body, so much will it retard the motion of the body, and draw it out of its straight course, independent of the resistance of the air, which will also act in opposition to my strength, and likewise retard the velocity of the body.

All motion in a straight line, is called projectile, therefore if I throw a ball up in the air, the law of motion is equally observed, as when I project it in a straight line horizontally, as the ball, in the former case, will ascend with a motion proportioned to the advantage which the force I exert has over gravity, and the resistance of the air.—We will suppose that the force I exert in throwing up a ball, to be such as would cause it to describe 60 feet in one second of time, did not gravity and the resistance of the air act in opposition to its ascent: but as the force of gravity, independent of the resistance of the air, causes all bodies to fall to the Earth at the rate of 16 feet nearly, in the first second of time, my ball would not rise above 44 feet in the first second. Nor will it rise those 44 feet in the next second of time, because in the next second of its ascent, the power of gravity increasing with the height, from which it falls, in the proportion of three to one, it will fall 48 feet; so that it will not actually rise above 12 feet. And as, in the third second, gravity will have the advantage over the projectile force, it will return and fall to the ground.—Let us suppose the ball as being projected from a piece of ordnance, in a horizontal position, although this is not the fact in practice; for were cannon planted in that direction, they never could discharge a ball to effect, as the whole force of the charge would be counteracted by gravity, as you will understand by the examples I am going to give you of projectiles; not that I propose instructing you in the ballistic art, that being foreign to my purpose; but having spoken of a ball being projected from a cannon in a horizontal direction, I fear, if I were not to undeceive you, you might suppose that to be the usual mode of placing cannon.

In continuing my illustration of the effects of motion and gravity, by a piece of ordnance, you are to suppose A, *fig. 3, plate* 8, to be a ball just discharged from a cannon, with a force capable of carrying it to G, in a straight line, if gravity did not act on it, which is an unnatural supposition : then we must explain how its activity is exerted, and in what degree it will retard, as well as prevent the ball from pursuing the straight line F G. The line from F to G is divided into four parts, which are to answer to four seconds of time; by which you will understand the curve that the ball will describe, in each second of its motion. In the first second, by the force impressed on it, it would describe the space from F to B; but gravity, acting in a perpendicular direction, will bring it down to C. In the next second of time (in which, if it had pursued its straight course, it would have arrived at Q,) the effect of gravity increasing in each second by the odd numbers, as the body approaches the Earth, will bring it down to D, which is three times as low as it was brought down by the same power, in the time of the first second of time. In the third second, it will be brought down five times as low as in the first second, as to E. And in the fourth second it will fall down to H.

These points form a curve, as the ball is gradually descending through the spaces FC, CD, DE, EH. This curve is called a

T

parabola, and in this direction all bodies fall, unless they are projected perpendicularly upwards, or directly downwards.

The curvilineal course of falling bodies projected straight forwards, is caused by the combined motion communicated to the body in a straight and oblique direction; the latter being the effect of gravity, which is uniformly accelerated; by which the former is uniformly counteracted.

In defining the theory of projectiles, by circumstances observed of things on our Earth, in order to your comprehending the motions of the heavenly bodies, I have not allowed for the resistance of the air, which, in practice on our Earth, makes the theory of projectiles more difficult, as, in fact, they do not agree with what I have advanced when the resistance of the air is taken in to the computation, the deviations from a parabola being in that case very considerable; but which are inapplicable to the motions of the heavenly bodies, which move in an unresisting medium; therefore the idea of the parabola perfectly accords with their motions, when the directions of gravity are considered as parallel to each other, but of an ellipse when the attraction tends always to one common centre.

Having observed to you, that bodies projected perpendicularly, fall in the same line of direction, you must perceive by the foregoing illustration, that their doing so, is occasioned by the projectile and gravitating force acting in the same line of direction.

Sir Isaac Newton's third law is, that re-action is always equal to, and in a contrary direction to action; and that this happens when both bodies are in motion, as well as when only one of them is impelled. From which we infer, that if one body presses another, either at rest or in motion, it is equally pressed by it; and if a body strikes another in motion, in a contrary direction, it communicates such part of its motion to that body in the same direction with itself, as it loses; and the velocity of each depends on the proportion of the bulk of the bodies; that which contains the greater quantity of matter, having less velocity than that which contains less. These being equally applicable to the laws of attractions, are adapted to the explanation of the motions of the heavenly bodies.

Motion has a variety of dependencies, which are duly estimated, according as they are equable, accelerated, or retarded. Equable motion is also resolved into simple and compound; compound motion into rectilinear and curvilinear.—Equable motion, is that by which a body describes equal spaces in equal times.—Accelerated motion, is that which is continually impelled, and thereby receives a continual augmentation of its velocity, which is the same as is effected by gravity.—Retarded motion is continually decreasing, in the same way that accelerated motion is continually increasing; and both are the effects of gravity or attraction, which, in the former, acting in opposition to the rectilineal motion, retards it; whereas, in the latter, by conspiring with that motion, it accelerates it.—Simple motion, is that which is given by one force only; but compound motion is produced by more than one force, which act upon the body in various directions.

Sir Isaac Newton, upon the principles of the motions of natural bodies, discovered that matter could not move itself, and therefore that a body in motion could not of itself change either its velocity or its direction; from which he was led to examine into the causes of those effects when produced. In order to estimate the proportion of those powers, it became necessary to measure them, which he was able to do by a due observation of their effects. That which makes a body heavy when at rest, accelerates its motion in falling, and bends its course into a curve. The force which causes a moving body to recede from a centre, or from the body round which it revolves, is called centrifugal force; and that which occasions it to approach a centre, is called centripetal.

The former is occasioned by the natural endeavour of the moving body to continue in its state of rectilineal motion, which, but for the latter, would carry it off in a tangent from the centre round which it revolves; and the latter, which urges a body towards a centre, and makes it revolve in a curve round it, is occasioned by gravity or attraction.

The centrifugal and centripetal forces require particular attention, being intimately connected with the motions of the heavenly bodies.

When a body describes an exact circle round a centre, we know that the centrifugal and centripetal forces are equal to each other, their action being always performed in a direct opposite order to each other.—But if we perceive a revolving body to recede farther from, and approach nearer to, the central body at one time, than at another, we are certain that in neither case can they be equal to each other; as, in the former, the centrifugal force must prevail, and in the latter, the centripetal. These positions you must bear in mind against I recall your attention to them, when treating of the planets orbits, the form of which depends on these circumstances.

I have informed you that the Sun is the largest body in the system, and that it is situated in the centre of all the other bodies of it; also, that those bodies being of different sizes, revolve in orbits exterior to the Sun, at different distances from this great central body. The certainty of this position will be evident to you, on considering the laws which govern the motions of revolving bodies.

First, Let us suppose two equal weights to be fixed to the two ends of a piece of wire, and the wire to be suspended on a point so as to turn round on it, when impelled forward; this circular motion could not be effected but by the point of suspension being placed at an equal distance from each weight, for the weights being equal to each other, the centre of gravity must fall directly at an equal distance from each, and round which alone it could revolve.

If, instead of two equal weights, two unequal ones were fixed to a wire; the point of suspension, or centre of gravity, will be removed nearer to the large weight than to the small one; and, therefore, when placed in a revolving situation, the smaller weight will describe a circle exterior to the larger one, because the centre of gravity will be further removed from it. The greater the disproportion of the weights the nearer will the centre of gravity be to one, and the further removed from the other; the centre of gravity being at that distance from each weight, which is in proportion to the difference of their gravity or attraction.

The Sun being the largest body of the Solar system, it must be nearer the centre of gravity than any other body belonging to it.

The Sun and the planets are connected together by attraction, and therefore are all actuated by the same impulse, only in a different degree; the motion of the Sun is very slow, and he describes but a very small circle round the centre of gravity; his motion being so very slow, and his bulk very considerable, we do not perceive that he moves. His motion on his axis is discoverable by the spots on his disk, and it is occasioned by his globular form and progressive motion.

I shall postpone further illustration till another opportunity, as I have furnished you with much to reflect upon; and as my future elucidations will be better understood by giving you time to exercise your minds, in establishing the principles I have already laid down; that being absolutely essential to your perfectly comprehending the affections of the heavenly bodies, which are all the result of the laws we have been considering; and which I have endeavoured to communicate in a familiar manner, and so as to fatigue you as little as possible, considering the dryness of the subject, which however unpleasing to you in description, will be found intrinsically valuable, by affording you much real satisfaction, as without these previous intimations you could not understand one half of the phœnomena I shall describe to you; but by a proper application of them, all will appear rational and easy, and you will be delighted by having the laws by which the Almighty conducts the affairs of the universe, thus rendered as familiarly pleasing to you, as they are infinitely sublime.

END OF THE SIXTH LECTURE.

Plate IX.

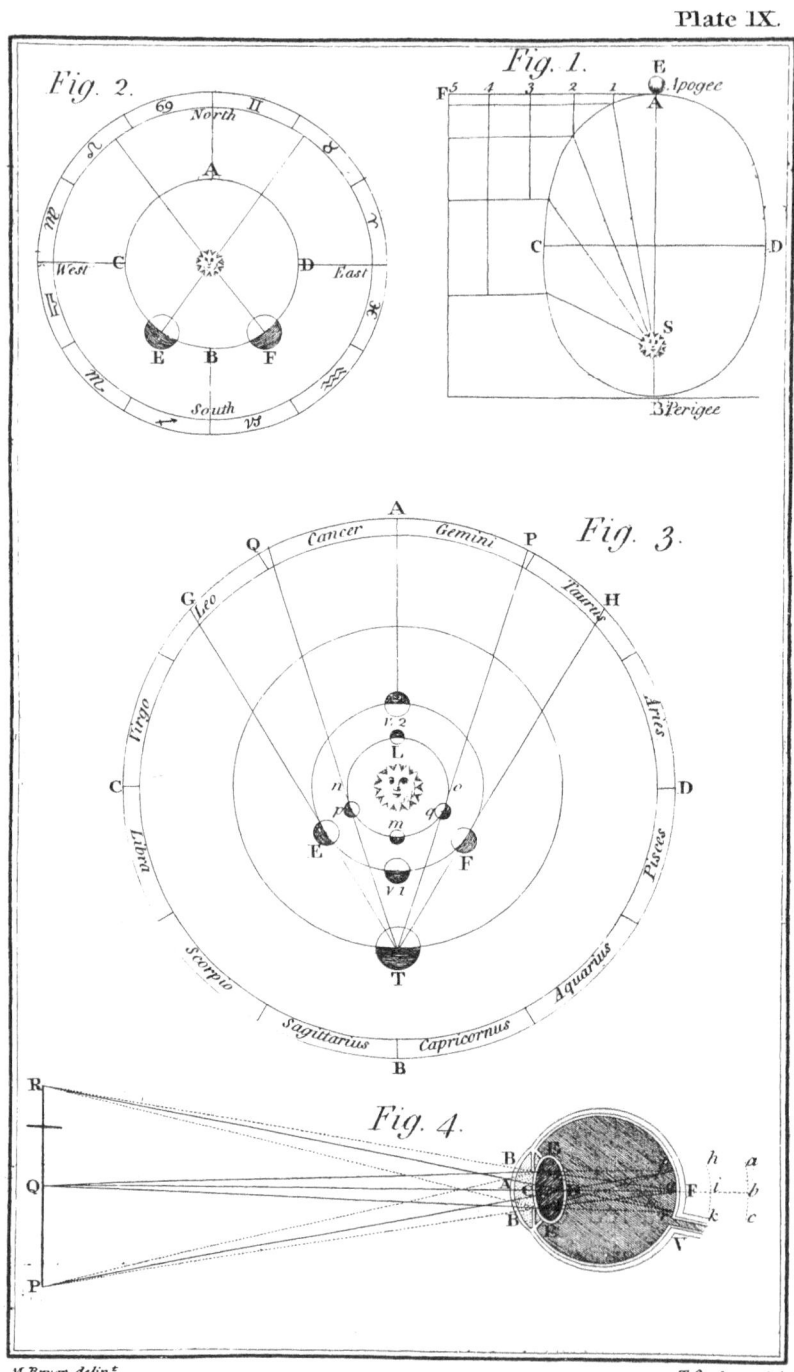

Fig. 2.

Fig. 1.

Fig. 3.

Fig. 4.

M.Bryan delint.

T. Conder sculpt.

LECTURE VII.

OF THE ORBITS OF THE PLANETS.—THE LAWS BY WHICH THE
PLANETARY MOTIONS ARE REGULATED.—OF THE NATURE
OF VISION, AND THE ORGANIZATION OF THE EYE.

> How charming is divine philosophy!
> Not harsh and crabbed, as dull fools suppose,
> But musical as is Apollo's lute,
> And a perpetual flow of nectar'd sweets.   MILTON.

How beautifully, how justly, has the sacred bard described the delights of philosophy! You have already tasted the sweet pleasures flowing from that never-failing spring; you will acquire a greater relish at every draught; drink deep and fear not — its freshness prevents surfeiting, and its variety precludes satiety.

Having examined gravity in general, I shall now apply it to the consideration of particular subjects, comprehended in its effects on the planets of our system.

The terms weight, gravity, attraction, and centripetal force, denote the same thing, only they are differently applied. In order to your understanding their proper application, I shall inform you how they are used, so as to be expressive of the different circumstances of the same thing.

When a body is impelled towards the Earth, the power by which it is drawn downwards, is called gravity.

When this power is considered as tending towards the centre of the Earth, it is called centripetal.

When a substance is supported in its way to the Earth, it is called weight; but when we refer it to the earth or planets motions in their orbits, we call it attraction.

Newton discovered that it was this power which retains the heavenly bodies in their orbits, and in their respective situations in regard to the Sun, and other bodies of the same system, according to their relative powers of attraction, as they all mutually attract each other.

By the aid of mathematics, the incomparable Newton ascertained the quantity of attraction necessary to make the planets revolve at certain distances from the Sun, and from each other, as also the satellites from their primaries.

We are certain that the planets must be retained in their orbits by some power which is continually acting upon them; that this power draws them towards the centre of their orbits; and that its effect increases as they approach the centre, and is diminished as they recede from it, in a certain proportion to the distances. The laws of this power have been found to be the same as those observed of gravity; and therefore we may believe their identity, which has been settled by imagining a line drawn from the Sun to either of the planets, when on observing the time taken by a planet in passing over an arch of the ellipsis, circumstances accorded exactly with what would happen in bodies falling towards our Earth by the power of gravity; i. e. the nearer the body was to the attracting power, or centre of the Sun, the quicker was its motion, and its motion was uniformly accelerated in approaching the Sun. The same has also been discovered of the Moon in respect to the Earth; the Earth gravitating towards the Moon, and the Moon towards the Earth, in proportion to their different powers of attraction, which occasion them to revolve round a centre of gravity between them, and to cause the Earth to be nearer to that centre than the Moon, because the Earth is the larger body.

The centripetal power of the primary planets is directed towards the centre of the Sun, and that of the satellites towards the centre of their primaries; which power is exerted in proportion to the distances of those bodies from their central bodies, being reciprocally as the squares of their distances from them. This evidently proves that it must be the same power actuating them all, and that this power is gravity or attraction—as the effects are the same with those observed in respect to bodies on our Earth.

The planets perform their revolutions round the Sun in eliptical orbits. Before I inform you why they do so, I shall explain the nature of an ellipse.

Fig. 4, *plate* 8, is an ellipse; its transverse axis, or longest diameter, A B; its shortest diameter C D; the two points F F the foci, which are so situated, that if any two right lines were drawn from them, and made to meet in a point in any part of the circumference of the ellipse, the sum of those two lines would be equal to the longer axis A B; as the two lines drawn from F F to G, *fig.* 4, are found to be, by applying a pair of compasses to the scale A B, *fig.* 5, *plate* 8.

You may say, perhaps, of what use is this illustration? as at first view its utility does not appear; but you will change your opinion when I inform you, that without comprehending the foci of an ellipse, you cannot understand why the Sun, being placed in the lower focus of every planet's orbit, should occasion him to be in the centre of gravity of all their mutual attractions, as at one time he must be further from them than at another, and from all the planets he must be at different distances—but by having the nature of the foci of an ellipse explained to you, all will become clearly understood.

Perhaps it may not be deemed intrusive if I introduce in this place the manner of forming an ellipse—by which instruction also my illustration of this part of the subject may be better understood.

Fix on a plane two pins, at any distance from each other, round which tye a loose thread according to the size of the ellipse you

U

wish to describe; then straining this thread with a pencil, carry it round with a steady hand, and the pencil will describe the curve line called an ellipse. Now, if there were but one fixed pin, the pencil at the extremity of the thread would, in being carried on, describe a complete circle; hence you conceive that the nearer the pins are to each other, the shorter will be the ellipse, and the nearer it will approach to the form of a circle.

The distance or space, F H, is called the eccentricity, which is different in the orbits of the different planets; but in all, it is so small that it is taken but little notice of. In the orbits of comets the eccentricity is very great, as, on account of their being very long ellipses, the foci are very distant from each other.

We are certain that the orbit of the Earth, as well as those of the other planets of our system, are ellipses, because the Sun appears of different sizes at different times of the year. His diameter in June is about 31′, and in December about 32′, on account of the sun being nearer to the Earth in winter than in summer.

The attraction between the Sun and the planets, and between the satellites and their primaries, must be mutual, and in proportion to their quantities of matter; for if it were not so, the Sun by his great attraction would draw all the bodies of our system into himself.

It is gravity or attraction which gives weight to bodies—forms all the planets into solid balls—circumscribes their orbits—and retains on their surfaces the beings created to inhabit them, notwithstanding the amazing velocity of their motions on their axes.— How great is the power, how universal the agency, of this invisible principle!

I have informed you, that if two forces are acting on a body in contrary directions, of unequal power, that the body impelled will move in the direction of the strongest force, and with a velocity equal to the difference of the velocities of the two forces. Thus, in reference to the motions of the heavenly bodies, we allow that they being impelled to move by divine command in a straight line, are brought down from that line by the power of gravity or attraction, and that they afterward move by the velocity gained in their fall; as we perceive things to do on our Earth.

The motions of the heavenly bodies are perpetual, because they do not move in a resisting medium, or suffer any friction, and because their influences are perpetual; yet their motions are accelerated and retarded according to the different effects of gravity acting upon them, as I shall further explain to you.

A body moving round the Sun has its motion accelerated in falling towards it, but is kept off from it by the centrifugal force acquired in falling, which force is independent of the Sun, and proceeding from the progressive motion of the impelled body, by the direction of which force alone it would go on in a straight line for ever; but the Sun drawing it from that line by its attraction or centripetal force, these two forces being exerted in contrary directions, instead of the body going off in a tangent to its orbit, by the force it has acquired in falling, it acquires a curvilinear motion; and it ascends again to the same height, because the power of attraction and the projectile force act upon it, both in its descent and ascent, only the effects are in a contrary order; but being in an equal degree to each other in both cases, the curve the body describes on each side of the Sun must be the same.

The Sun is situated in the lower focus of the elliptic orbit of every planet. The planes of the orbits of all the planets cut the Sun as it were in half, and are differently inclined to that of the Earth. Those two points of the orbits which cross that of the Earth, are called the nodes of the planets. That by which a planet goes above the plane of the Earth's orbit, is called its ascending node; and that by which it descends below the same, is called its descending node.

Nothing is more easy to be understood than the motions of the primary planets, being deducible from the known laws of two forces only—namely, the centrifugal, or force from the centre of an impelled body, produced by its natural tendency to proceed in a straight line, by the progressive impulse; and the centripetal, which causes two or more bodies in motion, connected together, to revolve round a centre; which when referred to the motions of the planets, the centrifugal force is known to perpetuate their motions, and the centripetal force to cause them to revolve round the Sun, by the superior attraction of that body. In regard to the effects of attraction between the planets, they are very inconsiderable, on account of their immense distance from each other, and comparatively small quantity of matter, although it is something when they are in the same quarter of the heavens, each attracting the other in proportion to their quantity of matter.

The motions of the satellites round their primaries, compounded with their motions round the common centre of gravity of the system, occasion them to be less uniform in their affections than the primaries in theirs, as you may suppose, on reflecting that each of them, although it gravitates towards its primary, yet it must also be affected by unequal attractions from the Sun, at different times, being sometimes nearer to that luminary, and at others further removed from it, than its mean distance.

Although the motions of the primary planets may be esteemed uniform, as far as respects their motions in the same parts of their orbits, being always at the same rate in each; yet are they not equal in all parts of them; on account of the Sun being placed in one of the foci of their elliptical orbits, and not in the centre.

When they are in that part of their orbits in which they are approaching the Sun, their motion becomes quicker and quicker; and when in that in which they are retiring from it, their motion is slower and slower. This I will further explain by a diagram.

Let the ellipse ABCD, *fig.* 1, *plate* 9, represent the orbit of a planet, S, the Sun, in the lower focus of the ellipse. Let AB (which is the longer axis of the ellipse) be the line of the apses,—the point at A the higher apsis, and that at B the lower.

The higher apsis is the place of a planet when it is at its greatest distance from the Sun, and which is also called its aphelion, or apogee; and the lower apsis is the place of a planet in its nearest approach to the Sun, and is called its perihelion, or perigee; also the line of the apses, or apsides, passes through the centre of the orbit of the planet, as well as through the centre of the Sun. The extremities of the line DC, *fig.* 1, *plate* 9, are the points of the planets orbits, at their mean distance from the Sun.

Suppose the ball at E to be a planet, which is projected in a straight direction towards F, through the etherial unresisting medium. If no other than this projectile force acted on it, it would go on for ever in that straight course, and thereby be deprived of its grand vivifying principle, the Sun; but the instant of its projection, the attraction of the Sun begins to affect it with a degree of power proportioned to its distance from it, and in a direction which acts perpendicularly to its projectile force; it is thereby brought down more and more from its straight course EF, in the proportion I have before shewn you by *fig.* 3, *plate* 8. This attraction acting more and more on the body as it approaches the Sun, forms the orbits of the planets into ellipses.

Thus the motion of a planet will be more and more accelerated in its way from A by C to B, *fig.* 1, *plate* 9, because the attracting power will be gaining continually on the projectile. But when it has arrived at B, it will have gained so much force from its centre, as to overcome the Sun's attraction, by which wise allotment of power the body is prevented approaching so near to the Sun as to be involved in its conflagration, which would totally annihilate it.

These two powers being duly weighed, or balanced, in the scale of Infinite Wisdom, prevents the projectile force from carrying the body off in a tangent from its orbit, in the direction B G, when it has arrived at B, by the Sun's attraction preserving it within certain limits. These two powers always acting on the body, cause it to describe the same curve in its departure from the Sun, that it described in its approach to it, by which it again arrives at A; the motion of the planet being as uniformly retarded by the Sun's attraction in its departure from that body, as it was accelerated in its approach to it. When the planet arrives at A, the Sun's attraction brings it down again in the same manner as before—and thus wonderfully are the motions of the planets perpetuated and circumscribed.

You now know why a planet's motion is swiftest when it is in its perihelion, and slowest at its aphelion. When it is at C or D, its motion is called mean, being between both extremes. These places of a planet are also called its mean distance from the Sun. The motion of a planet in these places of its orbit is such, as if continued always at the same rate, would cause it to describe its orbit exactly in the time it does describe it in, by means of the different velocities.

What I have advanced of the motions of the planets in their orbits, is applicable to those of the satellites, only there are some irregularities in the latter, the consequence of their obeying two central forces.

Such are the different causes of irregularity in the motion of the Moon, that it has but one strictly uniform, and that is on its axis; by which we always have nearly the same illuminated face of this satellite turned towards us, when it is visible.

I shall not at present speak of the irregularities and other affections of the Moon, as a better opportunity will offer for discussing this part of our subject, but proceed in applying what we know of motion and gravity, to the primary planets; having advanced sufficient of the laws of gravity and projectiles, for your comprehending their effects on the bodies of our system, of which I am about to treat.

There are, you know, seven planets belonging to the Sun, depending on his influences, already discovered, of which I shall speak. There may be more, but of these only we can treat confidently.

Each of these planets, of which our Earth is one, is encompassed by an atmosphere, by which the creatures inhabiting them are able to exist. We know that animal life cannot be continued without the impressions of the atmosphere of the Earth, and that its healthful state also depends on circumstances of it; therefore, it is natural to suppose, that all those bodies which are circumfused by that same fluid element, are supplied with it for the same beneficial and essential purposes. The motions of all the planets being also analogous to those of the Earth, we conclude that they are of the same nature, and thereby capable of supporting all animal and vegetable life.

Some of the other planets revolve nearer, and some at a greater distance from the Sun, than our Earth does, as I have pre-acquainted you; also that their periodic times are proportioned to their distances, and that these are all rationally accounted for on the theories of gravity and centrifugal force, and accord with them.

The mean distances of the planets are in the following proportion, which proportion is represented by the circular orbits of the planets delineated *plate 6*. The distance of the orbit of the Georgium Sidus from the Sun, is divided into 190 parts, as is exhibited on the scale drawn from the Sun at A, to the orbit of that planet at B. Saturn is situated at 95 of those parts, Jupiter at 52, Mars at 15, the Earth at 10, Venus at 7, and Mercury at 4.

The annual motions of the planets are ascertained by their change of place in the heavens; being removed from one place in the zodiac to another. Their direct motion is according to the order of the signs, from west to east; and this is their true motion; their revolution on their axis is performed in the like direction.

The apparent diurnal motion of all the heavenly bodies above our horizon, from east to west, is occasioned by the diurnal motion of our Earth, which corresponds with that of all the other planets, being performed from west to east.

The planets, as viewed from our Earth, have another apparent motion, on account of the annual motion of the Earth and them, in the same direction, with different velocities. This motion is called retrograde, as it makes them appear to us to go back again, as it were, by transfering their places in the heavens to different situations, in respect to the Sun and us, as from the east to the west side of that luminary, at particular times. The cause of this deception I shall, in due time, fully explain to you, both by diagrams, and the instrument called a planetarium.

All the planets and comets of our system appear bright, by reflecting the light of the Sun, as is known by the different phases of those (which are visible to us) at different parts of their orbits.

Those planets which, in revolving round the Sun, regard him as the centre of their motions, are called primary; and those which regard a primary planet, as the centre of their motion, revolving round the centre of gravity between it and themselves, are called secondary planets, or satellites.

Comets are primaries, because they regard the Sun, or centre of gravity between them and the Sun, as the centre of their motion.

Such of the primaries as are situated most remotely from the Sun, have their quantity of reflected light considerably augmented by the number of Moons, or satellites, which accompany them in their revolutions.

The satellites also appear to be habitable worlds, if we may judge of the rest, by what we perceive of that which belongs to our Earth, the vicinity of which to us, renders the mountains and cavities on its surface visible, by the aid of telescopes.

The Earth not being placed in the centre of the periodic motions of the planets, they must be sometimes nearer to us than at others, and of consequence appear larger or smaller, in proportion to their distances from us, as the apparent sizes of bodies diminish, according as their distances increase ; because it is not their real diameter that we can judge of by appearances, but the angles under which they are viewed, which angles will be greater or less, in proportion to their distances. For the foregoing reason also, at a given distance, bodies which differ in size will also differ in apparent magnitude. If a very large body be placed at as much greater a distance from us, as it exceeds in size a less body in a nearer situation, they will both have the same apparent magnitude or diameter. The position of an object will alter its apparent diameter, as will also the degree of light under which it is viewed ;—the latter effect is produced by the different extensions of the pupil of the eye in different degrees of light. If we look at an object in a very strong light, the pupil contracts ; but if in a very weak light, it expands, and of course, in the former case, the angle under which the body is seen, at a given distance, must be diminished, but in the latter it must be enlarged ; and therefore the apparent size of the body must agree with these circumstances. For the above reasons, all objects viewed from our atmosphere, in a medium more dense or obscure than it, appear larger than they do in the air.

I feel an irresistible impulse, which I trust I shall be excused for obeying, to introduce in this place, some observations on the wonderful construction of the eye, the admirable disposition of its parts, and the happy and nice powers of adjustment they are endowed with, to answer all the purposes of vision.

The eye is the instrument invented by Infinite Wisdom, to con-

X

vey to our mind the idea of visible objects—to doubt of the truth of this assertion, is impossible, when we examine its structure, and the effects produced on it by the rays of light.

The eye is placed in a bony cavity, called the orbit; its form is globular; within this globe are contained three different kinds of humors, enclosed in several distinct sorts of teguments, or coats, in which blood vessels, nerves, and arteries, are curiously interwoven.

I shall first treat of the external advantages attendant on the mechanism of the eye, and its concomitant appendages.

The inside of the orbit which contains the eye, is lined with a lubricating and membraneous substance, which affords the eye a soft bed to perform its movements in, without injury to its delicate substance. Those arches of hair, called eye-brows and eye-lids, are not less useful than beautiful; for they defend the eyes from too strong a light, and prevent dust or other small substances from falling into them, by being provided with muscles, for the purpose of projecting or drawing them down, so as to defend the eye from a glare of light, and from incumbent particles of dust.

The eye-lids afford also a perfect and secure asylum for the eye when we sleep, or have occasion to guard against external injury.— When we are awake, and unfearing of external annoyance, the eye-lids, by their motion, diffuse a fluid over the eye, which keeps it constantly moist and clear, by which alone it could answer the purposes of vision.

These lids join at their two extremities, and that they may shut with greater exactness, and not fall into wrinkles when they are elevated, each edge is stiffened with a cartilaginous arch, which is bordered with hair; by the latter the contour of the eye-lids is softened, the eye protected from straggling motes, and the light moderated in its approach to the retina. The eye-lids also assist in these desirable effects, by excluding a superabundant quantity of light.

The upper part of the orbit of the eye has a gland placed in it, which constantly furnishes sufficient moisture for keeping the part of the eye exposed to the air, in a proper state of lubricity and pellucidity; and that this purpose may be fully answered without our attention to it, we shut the eye-lids, or wink our eyes, without the concurrence of our will or reason.

The corner of the eye next the nose, is provided with a curuncle, for the purpose of keeping that corner of the eye from being perfectly closed, that any tears, &c. may flow from under the eye-lids, when we sleep, into two little holes, one of which is in each eye-lid near the corner, for carrying off any superfluous moisture.

The eye is furnished with six muscles, which spread their tendons far over the eye, in order to effect a motion in every direction, excepting an oblique one towards the nose, which is aided by a particular auxiliary.—The side of the eye next to the nose not allowing room for a muscle, a small bone is placed on the side of the nose, with a hole in it, which serves as a pulley for the tendon of a muscle to pass through, by which an oblique direction of the eye is obtained.

The eyes have a parallel or uniform motion, in which they always coincide; this is extraordinary to human reason, as the organs of the two eyes are totally distinct, having no communication with each other, and yet they appear actuated by the same force or mechanism. The purposes supposed to be effected by this union of action and direction, is that of seeing things single, which are viewed double. I shall give you Sir Isaac Newton's supposition respecting our seeing things single, which are painted double, that is, the two images of the object painted on the retina, one in each eye, appearing but as one to the imagination.

" The species of objects seen with both eyes, may unite where the optic nerves meet before they come into the brain, the fibres of both nerves uniting there, and after union, going thence into the brain, in the nerve which is on the right side of the head, and the fibres on the left side of both nerves uniting in the same place, and after union going into the brain in the nerve which is on the left side of the head; and these two nerves meeting in the brain in such a manner, that their fibres make but one intire species or picture, half of which is on the right side of the sensorium, and comes from the right side of both eyes through the right side of both optic nerves, to the place where the nerves meet, and from thence to the right side of the head into the brain; and the other half on the left side, comes in the like manner from the left side of both eyes. The optic nerves of all animals which look the same way, as men, horses, dogs, &c. meet before they come into the brain, but the optic nerves of such animals as do not look the same way with both eyes, as fishes, do not meet before they go into the brain."

This conjecture appears reasonable, and may therefore be admitted; but the effect must after all be referred to the mind, as well as what causes that involuntary motion which produces the effect, or that motion which causes the image to be seen at all; for although undoubtedly vision, or the appearance of objects, is occasioned by the pictures on the retina, yet the eye can see no part of itself—it is the mind that perceives and judges; the eye is only the medium, or instrument, by which the idea is conveyed to the mind, and for the operations of the mind upon the body, or the body upon the mind, we are unable to account.

Considering the eye merely as an instrument, we need not enquire why, when the pictures of objects are painted in it, in a reverse posture, our imagination perceives them upright; to solve which difficulty, anatomists have been unable; nor can they ever afford us a rational solution of a circumstance independent of the organization of the human body. All our senses are aided by the mechanism of the organs created for their use, but their impressions are referable only to the spirit—the understanding; and therefore not definable by human comprehension.

Having treated of the principal external parts of the eye, and the advantages procured by their nice adjustment, I shall venture to speak of its internal parts; which will be less digressive, as all the instruments used to aid astronomical investigation have been constructed upon the principles of refraction, reflection, &c. effected by the various humors and coats of the eye, and therefore they will be better understood from a description of this grand original and its affections.

The globe of the eye consists of several coats, containing three pellucid liquors, which are so adjusted that the rays proceeding from luminous objects, and admitted at the fore part of the eye, called the pupil, are brought to a focus on the back part of it.

The outer coat, or sclerotica, is a hard substance of a whitish colour, resembling parchment, the hinder part of which is very thick, and is opake; from whence it becomes gradually thinner and thinner as it approaches the part in front of the eye, where the white terminates; the other part of this tegument is thin and transparent, and projects a little, forming the segment of a smaller sphere; this part is called the cornea, from its transparency; this quality of it is necessary for the free admission of the light.— This membrane is composed of several layers, and replenished with clear water and pellucid vessels.

The second coat of the eye, or the choroides, is soft and tender, is composed of innumerable little vessels, and it adheres to the sclerotica; it is outwardly of a brown colour, and inwardly almost black. This tegument, like the sclerotica, is distinguished by two names, the fore part being called the uvea, and the hind part the choroides.

The fore part, or uvea, commences where the cornea begins, i. e. at the edge of that dark part of the eye called the iris. It is attached to the sclerotica by a narrow circular rim, from which part the choroides divides from the sclerotica, or changes its direction, turning inwards to the axis of the eye, or middle of the globe of the eye, and thus forms that round hole we call the pupil. The uvea commences where the choroides divides from the sclerotica, from which part to where it turns inwards and forms the hole called the pupil, is called the iris, which is composed of the dark colour of the choroides, called the uvea, combined with the reflections of the light occasioned by the puckering of the membrane on turning inwards.

The pupil of the eye has no determinate size, but depends on the action of the membrane which forms it, which either expands or contracts it, so as to accommodate the organ of sight to the strongest or weakest impressions of the particles of light; as thus: When the light is too intense, the pupil is contracted, to prevent the admission of too great a quantity of light, which would injure the sight; but when the light is weaker, the pupil is enlarged, and thereby a greater quantity of the rays of light fall upon the retina, in order to render it in both cases duly active. The whole of the choroides is opake, therefore no light can enter the eye but what passes through the pupil.

The third and last membrane of the eye is called the retina, because it is spread like a net over the bottom of the eye. It is a continuation of the optic nerve, and lines the inside of the choroides; and the concave side of it covers the surface of the vitreous humor, terminating where the choroides turn inwards, so that it contains the vitreous humor. On this membrane within-side the eye, that is, on its convex surface, are painted the images of objects.

The coats contain the humors of the eye; one humor forms a solid substance, another is soft, and the other perfectly liquid. The humors are of such forms and transparency as are best adapted

for transmitting the rays of light, and placing them in positions favorable to distinct vision.—They are all clear like pure water, possessing no essential colouring particles; therefore the colours exhibited by them, must be derived from the impressions of the different particles of light.

The most fluid of these humors is called aqueous; it fills the
interstice between the cornea and the pupil, and also the space between the latter and the chrystaline humor. Its form is plano-
convex, its quantity is so abundant, that it swells out the fore
part of the eye into the segment of a small sphere. It is not
known from whence this humor is supplied, yet its source is so un-
failing, that if the coat which contains it be wounded, so that the
humor all flows out, if the eye is kept closed a proper time for
the wound to heal, the fluid will be recruited.

The second humor is called the crystalline; it is as transparent as
the aqueous, but less in quantity, and more dense, of the consist-
ency of a stiff jelly. Its form is doubly convex, but the two parts
are of different convexities; the most convex part is received into
an equal concavity in the vitreous humor.

The crystaline is contained in a kind of case, the fore part of
which is thick and elastic, the hind part thin and soft. This case
is suspended in its place by a muscle, which, together with the
crystaline, divides the globe of the eye into two unequal portions,
the smaller and foremost containing the aqueous humor, the larger
and posterior the vitreous.

The crystaline humor has no visible communication with its case,
for when it is opened the humor slips out.

In old age the crystaline becomes discoloured, and therefore all
objects appear less bright, and are tinged with yellow, which de-
generacy, not coming on suddenly, prevents old people from being
sensible of this effect of it.

The vitreous is the third and last humor of the eye, and appears like glass; it is neither so dense as the crystaline, nor so liquid as the aqueous. It fills the greatest part of the globe of the eye, filling all the space between the sclerotica, from the insertion of the optic nerve to the crystaline lens.

The optic nerve passes out of the seat of the brain through a small hole in the bottom of the orbit of the eye; it enters the orbit of a form nearly globular, but compressed, and is inserted into the globe of the eye nearly in the middle, though not quite so, but rather higher and nearer to the nose.

Having said sufficient of the construction of the organ of sight for your comprehending the wise and wonderful disposition of its parts, I shall proceed to instruct you in the mode of its administration, and convince you that the eye was made for the rays of light, so as to enable you to confute the ignorant and unreflecting.

Fig. 4, *plate* 9, represents the coats and humors of the eye. The outer coat of the eye, called the sclerotica, is represented by the exterior of the lighter shaded circles B F B; the more convex part of which, from B to B, is the cornea.

The choroides is represented by the inner light-shaded circle B F B.

United with the choroides, and a part of them, is the uvea, which reaches from B B to *a a*.

Between A A is the pupil.

V is the optic nerve which spreads itself within the choroides, and in the figure it is represented by the dark part of it inclosed in the circles B F B.

E E is the crystaline humor.

B A B *b* C *b* contains the aqueous humor.

And the large space B E D E B F the vitreous.

Vision, as far as the eye is concerned, is effected by such refraction of the rays of light, by the skins and humors of the eye, as

causes distinct images of exterior objects to be painted therein—
and which purpose they are admirably contrived for effecting.

All the rays of light which enter the pupil from any luminous
object, are united closely together upon the retina, by which they
make a stronger impression than in their simple state. The retina is placed at a proper distance from the refracting substances,
in order that each pencil of rays may be received on it in distinct
focusses so that the images may appear conspicuous.

The degrees of the power of refraction of each humor of the
eye have not been determined for all states of it; but those of the
aqueous and vitreous humors, in their sound and perfect state, are
supposed to be much the same with common water, and that of
the crystaline a little more.

The humors of the eye altogether form a convex lens, the effects
of which will be easily understood by the diagram.

Let P Q R, *fig.* 4, *plate* 9, be a luminous object. The pencils
of light B P B, B Q B, B R B, from the points P, Q, R, falling upon
the cornea B A B, are refracted so as that they would belong to
focusses beyond the eye, at *a, b, c*; but the surface of the crystaline humor, C, increasing the degree of the refraction, would make
them meet in focusses nearer to the eye, but yet exterior to it, at
h, i, k. But lastly, the refraction being still augmented, by their
passing out of the crystaline humor into the vitreous, they are
brought to their proper focusses on the retina, at *p, q, r.* You will
recollect that the rays of light, in falling upon a lens of unequal
convexities, are refracted in passing in and out of that lens, in proportion to the different convexities of the two surfaces; and in a
contrary direction. Also, that the refracted rays meet in a point,
at a certain distance from the refracting surface they depart from;
and that as the rays of light, in passing through a convex lens, are
each differently refracted, they will meet in different points, but
at the same distance from the refracting surface.

Y

This illustration of the nature and effects of vision will not, I hope, be thought intrusive, as the subject is truly curious and interesting; and as many who may peruse this Book, may not have considered these effects, or even the nature of the organ of sight, so as to derive the many consolations—the many delights, both of body and mind, offered in this epitome of Infinite Wisdom and Benevolence.

Thus, whether we soar in contemplation of the majesty and glory of God displayed in the Heavens, or pursue our scrutiny of the wonders and benevolence of his administration and dispensations, manifested in the organization and effects of things upon Earth,—still we discover new cause for congratulation, new sources of delight and adoration.

Can any thing be more delightful than employing our understandings in researches, which bring us nearer in thought to God, and which convey to us such sure testimonies of his love?

When, in contemplating the powers the Almighty employs in regulating the planetary system, we perceive that all nature is supported by two opposite forces, acting upon each other, by which an harmonious equilibrium universally prevails,—how do we admire the greatness of that power, the infallibility of that judgment —that can thus appropriate and adjust properties to the effects required!

When we investigate the organization of the objects of nature, and perceive that all the wisdom and power of God is employed for the benefit of his creatures, how do our hearts glow with love and thankfulness—the sweet sensations it produces diffuse a serenity, which, like a charm, buries our cares in oblivion.

Let us endeavour to copy the attributes of this bright original, by using our power and abilities, not to afflict or be useless to our fellow-creatures—but to protect and to serve them.

Our understanding was given us in trust, like all other blessings;

for the use and abuse of which, we must one day give an account. Therefore, independent of ten thousand inconveniences, which result from not employing this great boon to the honor of God, we shall also incur the displeasure of that Divine Dispenser of good gifts, who expects that we shall improve every talent committed to our charge.

If you recur to the numerous instances in history, of the abuse and neglect I am speaking of, you will perceive the baneful effects of them in those who, not using their understanding aright, or by neglect in the cultivation of it, have become grossly depraved, and cruel in the extreme : their minds becoming impure and hardened, they lost all relish for what was lovely and good, and delighted only in all that was evil, till at last they became despised and abhorred of mankind. It is unnecessary in this place to enumerate those instances of depravity, evinced in the conduct of those who have neglected the cultivation of virtuous, benevolent, and religious sentiments, in order to deter my pupils from falling into the same errors, as they are familiarly acquainted with them in their course of reading and digesting the history of past ages ; and as they have testified their abhorrence, by shuddering at the horrid representation of those crimes, and have gladly turned from such unworthy and ungenial subjects of our animadversion, to those, in which the human character has been delineated with all its attractive charms of sensibility and integrity, and in which the most refined and highly cultivated understandings have been pourtrayed.

The contrasted effects of a virtuous and philanthropic, to those of a vicious and hardened disposition, strikes every sensible mind most forcibly. Let us then endeavour to avoid what leads to the one, and cherish all that perfects the other, as no one is wicked or hardened all at once ; and as virtue, like the security of all else that is truly valuable, requires vigilance.

No subjects tend more to invite us to the exercises of all religious and social duties, than those of Astronomy and Natural Philosophy.

Let me entreat you, my dear pupils, to imitate the harmonious effusions of nature, by being uniformly kind to each other, and by conducting yourselves with innocency and truth, so fulfilling your appointment upon Earth, and with all simplicity effecting the purposes of your being. Then shall the virtuous, the benevolent impressions you have received, be reflected in your actions, in as charming a manner as the light of the Sun from the bosom of the rose; and your virtues be as communicable, as delightful, in their influences, as the sweet effluvia of that flower diffused by the circumfluent atmosphere.

END OF THE SEVENTH LECTURE.

Plate X.

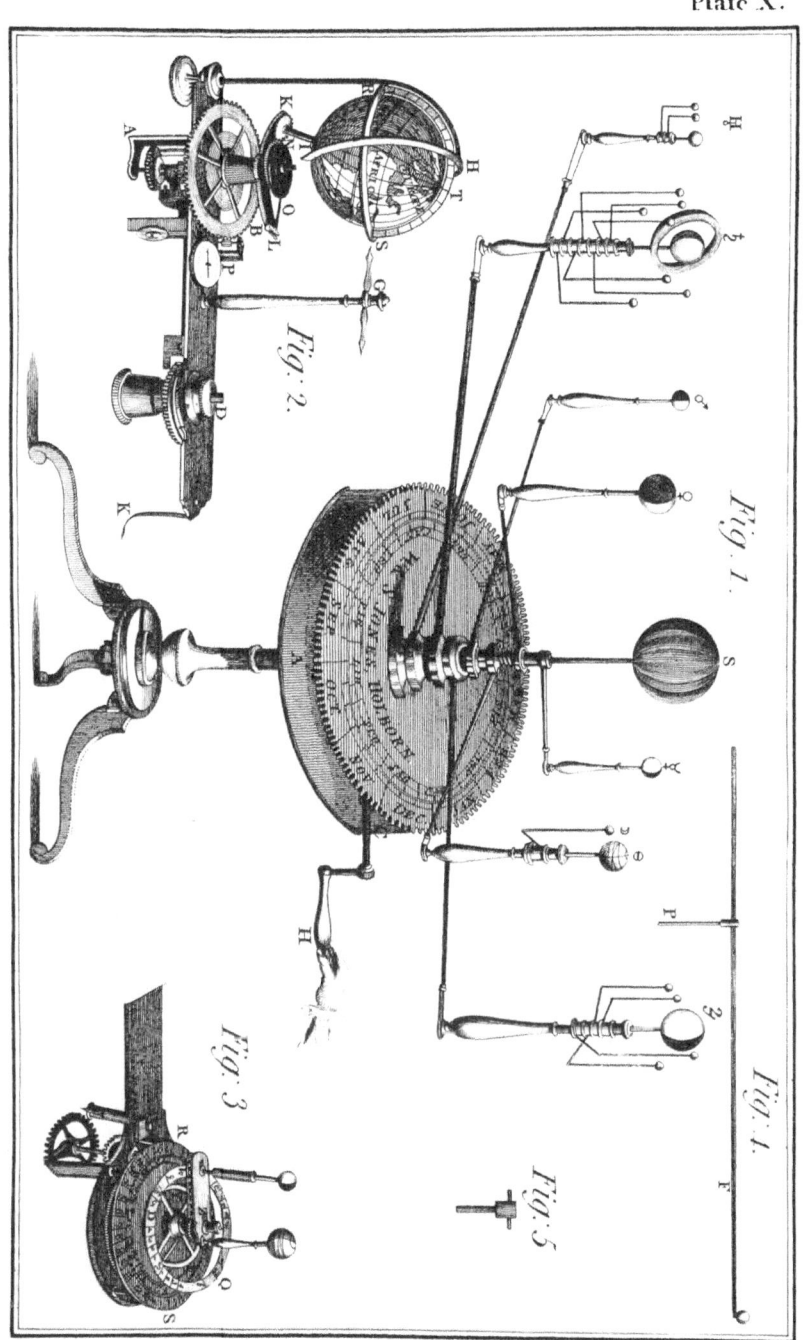

Fig. 2.

Fig. 1.

Fig. 4.

Fig. 3.

Fig. 5.

T. Conder sculp.

Plate XI.

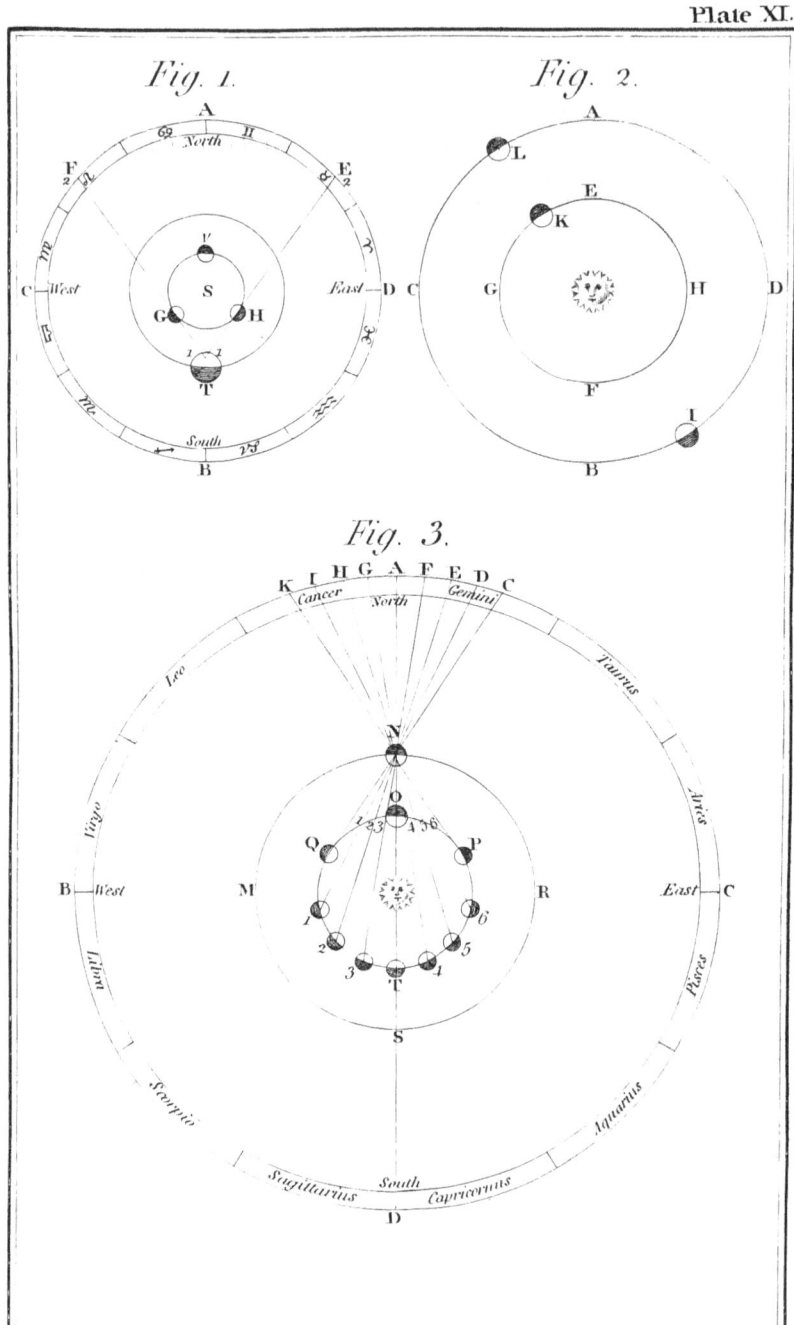

Fig. 1.

Fig. 2.

Fig. 3.

M.Bryan delin.t   T.Conder sculp.t

LECTURE VIII.

A DESCRIPTION OF THE ORRERY IN ITS DIVISIBLE STATE, AND
THE MOTIONS OF THE PLANETS OF OUR SYSTEM, ELUCIDATED
BY THE APPLICATION OF THE PLANETARIUM PART OF
IT; ALSO BY DIAGRAMS AND THE KNOWN LAWS OF
PROJECTILE AND CENTRIPETAL FORCES.

THIS instrument was constructed by Mr. Martin, and is now
sold by Messrs. Jones, Holborn. It is called a planetarium, and
is represented, together with its appendages, *plate* 10. The whole
forms a complete and elegant construction of an orrery; and which
being divided into its respective particulars of elucidation, namely
of the planets, the seasons, and the Moon, renders it preferable to
the largest compound orrery, in which the whole phœnomena are
exhibited at the same time, such an assemblage of movements rather causing confusion in the spectacle, unfavorable to accurate and
individual observation.

In *fig.* 1, *plate* 10, the instrument is furnished for the purpose
of elucidating the planetary phœnomena, and is in this arrangement of it called a planetarium.

The seven primary planets are placed in their proper stations,
and accompanied by their respective satellites. They are all put
in motion by turning the handle H, which communicates with the
wheel-work in the box A.

This machine illustrates the annual motion of the planets, on
which depend many curious phœnomena, one of which is the apparent retrogression, which the parts 4 and 5 are particularly calculated to exhibit.

Fig. 2 represents the appendage, which when attached to the plate of the orrery, forms a tellurium, for which purpose the planets must be taken off the supporting stem in the centre of the instrument, and *fig.* 2 placed on it, and fastened down by a screw at *a*. Under the globe of the Earth, *fig.* 2, is an index N O, which shews syderial time, or that in which the Earth performs its diurnal rotation. At P is another index, which shews solar time, or that in which the Sun transits the meridian of any place on our Earth. H and I is called the terminate, because it divides the dark from the illuminated part of the Earth. G is the solar ray, shewing to what part of the Earth the Sun is vertical, for every day in the year, as the Earth revolves round the Sun, and on its axis; also its declination.

The arm, K, rests on the plate of the orrery, and serves to point out the Sun's place in the ecliptic circle (which is represented on the outer circle of the plate of the orrery) as transferred by the annual revolution of our Earth in its orbit. R S T are moveable circles, which shew the longitude and latitude of every place on our Earth, when the circle R S is placed over the equator.

These are sometimes called the moveable horizon and meridian; the former is used with strict propriety, but not the latter when referred to this instrument, in which the terminator H I always represents the visible horizon, and therefore the circle R S is there properly a circle of longitude only.

Fig. 3, represents that part of the orrery which is adapted to illustrate the phœnomena of the Moon, and when it is placed on the plate of the orrery, the instrument is called a lunarium.

Having given you a general description of this curious orrery, I shall proceed to employ it in its individual arrangement, and first render it a planetarium.

The planets are placed on the supporting stem, according to their respective situations in regard to the Sun, which latter is represented by a large brass ball S. The signs of the zodiac are engraven on the planesphere according to their apparent arrangement in the heavens, to which the places of the Sun and planets are transferred. The signs have their respective or correspondent months and days also engraven on a circle exterior to them; so that the sun and planets may be placed in their proper situation for every day in the year, referring to White's Ephemeris, which informs us of the situation of each, with their aspects in respect to our Earth and the Sun.

The assumed period and orbits of the three most remarkable comets are represented on the plate of the orrery, as well as the known periods of Jupiter, Saturn, and the Georgium Sidus, and their orbits.

The planets are represented by ivory balls; the one half of each of them black, and the other half white; the white or illuminated sides of the planets being always turned towards the Sun; their known phases, as seen by a spectator on the Earth, is represented by the revolutions of the planets, which latter is effected by turning the handle of the instrument, which puts each planet in motion round the Sun, and with the different degrees of velocity they actually have in the heavens, so that their aspects in respect to the Earth are seen very clearly; and as the planets move independently of each other, they perform their periods in the same relative time as the real planets do in their orbits.

On putting the machine in motion we must observe the place of any planet in the ecliptic; and when it returns to the place from whence it started, see how many days it has been in performing its revolution, and we shall find that it corresponds exactly with the known period of that planet.

To observe their relative velocities, we must take the Earth as a standard, and we shall find that they all observe the same relative velocities as they really have in the heavens.

Suppose we place Mercury and our Earth on the same point of the ecliptic; when Mercury arrives at the same part from which he started, we shall perceive that the Earth has advanced only 87 days in her annual orbit. In like manner we may observe the periods and relative velocities of each planet, by taking our Earth and time as a standard.

Venus and Mercury are called inferior planets, because their orbits are included within the Earth's, and of course their revolutions are performed in less time; that of Mercury being performed in 87 of our days, and that of Venus in 224.

The revolutions of the planets in this instrument are to be expressed in round numbers, the nicer divisions of time not being allowed for in the planetarium. In pursuing our observations, we shall perceive that Mars performs his revolution on this instrument in 686 days, Jupiter in 4332, Saturn in 10761, and the Georgium Sidus in 30445; all of which accord with their known periods, excepting the difference of hours and minutes. Those planets which describe orbits exterior to that of the Earth, are called superior; the motion of the most remote planet is so slow, as scarcely to be observed, but by its effects, on this instrument.

The little white balls represent the Moons or satellites which accompany the planets in their revolutions; which, being stationary, excepting as placed by the hand on the instrument, have not their illuminated sides distinguished from the other, for they cannot exhibit their proper phases, as they do not describe their orbits round their central bodies, but only round the Sun. Another part of this machine is adapted to exhibit all the phœnomena of our Moon, which is the most essential to us to be informed of, as the phases of the others may be inferred from their known position, as set down in the Ephemeris, which is all that is necessary for our observation of them. The primary planets of this instrument have no diurnal revolution, those also being unnecessary, as *fig. 2, plate* 10, is constructed to illustrate all that respects the diurnal motion of the Earth very clearly; and of the others, we know that they have such a motion, which is quite sufficient, as of most of them we have ocular demonstration, by the different situation of certain spots on their disks, these moving from one side of the planet to another, as seen from our Earth, and disappearing and reappearing; also the motion of these spots being always uniformly performed in the same way, going off from the western edge, and reappearing on the eastern.—The distance between the spots appearing narrower at the margin than when at the middle of the disk, proves these bodies to be globular, and that they turn on their axes.

All loose or soft bodies turning on an axis, must become oblate spheroids, which is found to be true, in respect to the Earth and several of the planets. All globes which do not turn on their axes, are perfect spheres, on account of the equality of weight. The former obtain a spheroidical form, by their equatorial parts moving quickest, by which velocity they have a greater tendency to fly off from the centre in a rectilinear direction, which raises the middle or equatorial parts highest.

All the heavenly bodies appear at an equal distance from us, although their real distances are very different.

The Moon is nearer to us than any other body in our system, yet she appears at the same distance from us as the Sun does, and all the heavenly bodies appear as if placed in a concave sphere; which deception arises from our eye being in the centre of the sphere of light of their illumination; and therefore we cannot properly judge of their distance, or that of any object placed beyond the ordinary reach or circle of our view, having no means of judging of objects without that sphere, as we cannot compare them with intervening ones, as we can of objects within that sphere. That this is the cause of the deception alluded to, is evident, even

Z

by luminous objects on our Earth in a dark night; when we can see no intervening object, by reason of the natural sphere of light being lost in obscurity. As thus.—In approaching a town, in which we cannot discern any thing but the lamps; the buildings or intervening objects not being discerned by us, we cannot say which of the lights is nearest to us, nor can we judge of the real distance of any of them; and thus it is with celestial objects.

We may conclude that the fixed stars appear of different magnitudes to us, not so much from any difference in their sizes, but by being placed at different distances from our Earth, as their different times of appearing after sun-set seem to determine; for after being accustomed to the bright light of day for many hours, it is most probable that those stars which, by their situation, appear the brightest to us, must be first seen, on account of their superior brilliancy, occasioned by their situation in respect to the Sun and Earth, and not on account of their real size; for we know that when a room is first darkened, a small glimmering light is not perceived, although a bright light of the same size would be; but that when the eye has been accustomed to the gloom of the apartment, it becomes sensible to the least light admitted through the smallest crevice of it.

When the Sun has sunk 12 degrees below the horizon, the stars denominated of the first magnitude appear; and these we may suppose to be the nearest to us, both on account of their apparent magnitude, and also by their being first perceived by us. When the Sun has sunk 13 degrees below the horizon, stars of the second magnitude and degree of lustre are perceived; when 14 degrees below it, those of the third degree of both; when 15 degrees, those of the fourth degree; and when 18 degrees, those of the fifth and sixth degree of lustre and apparent magnitude; these are all that are visible without the aid of glasses; those perceived by their assistance are, on that account, called telescopic stars.

Those stars which Astronomers have not reduced to classes or magnitudes, are called nebulæ or light clouds.

The most remarkable star in each constellation, is marked with the first letter of the Greek alphabet; the next with the second letter in that alphabet, and so on. When there are more stars in the constellation than there are letters in the Greek alphabet, the superabundant ones are distinguished by the Roman characters a, b, c, &c. and when there have not been sufficient to express them all, the numeral figures have been used 1, 2, 3, &c.

Let us now resume the consideration of the bodies which compose the Solar system. You know that all the planets, and their satellites, shine by reflecting the light of the Sun; you are therefore sensible that the side turned towards that luminary, must be light; and as the planets are opake bodies, the side turned from the Sun must be involved in darkness, and cast a shadow from them on the opposite side, by intercepting the rays of light.

Sir Isaac Newton has calculated the quantity of heat in the Sun by the bulk of that body, and by its heat at the surface of the Earth; also its proportional heat at all the other planets, excepting the Georgium Sidus.

He says, that at Mercury the heat and light from the Sun must be near seven times as much as it is with us, on account of the comparative sizes of the Earth and Sun with that of Mercury and the Sun, and their respective quantities of matter and distances from each other.

In like manner he has also determined the degrees of heat and light received from the Sun by each planet belonging to the system: Venus receiving nearly twice as much as the Earth receives, Mars not half as much as the Earth, Jupiter not one-thirtieth part so much as the Earth, and Saturn about one-hundredth part of what is received by the Earth. His mode of calculation may be understood, by applying it to one planet only; as thus—Having found the distance of Saturn from the Earth and the Sun, and its real diameter by its apparent one, as seen from the Earth; by solid geometry he discovered that the Sun contained 2360 times as much matter as Saturn, and consequently, by comparing the known heat at the distance from the Sun at which our Earth is placed, he could ascertain the heat at Saturn. Also, by knowing the sizes of each planet, and the Sun's power on them, he could estimate the distance of the centre of gravity between each planet and the Sun; so that of Jupiter he found to be nearly in the superfices of the Sun; that of Saturn a little within it; and that the common centre of gravity of all the planets with the Sun, is not further from that luminary than his semi-diameter, as I have before informed you; also, that round this centre the Sun revolves, by the united attractions of all the planets, combined with his own.

In this instrument, *fig.* 1, *plate* 10, the orbits of all the planets are perfectly round, and in one plane, which is not exactly according to fact, as they acquire an eliptical form by the observance of two motions; they also cross each other, and the plane of the Sun's orbit in different parts of their orbits, and with different inclinations to it. But these circumstances, in which this instrument differs from the true shape and inclination of the planets orbits, is no impediment to the application of it, in regard to the phœnomena of the places of the planets in their orbits; and any deficiency respecting the nodes, the Ephemeris will always supply for any time, or any planet, shewing the place of the nodes; and if we want to distinguish the exact point of the Heavens in which any planet will appear at any given time, we need only make use of a quadrant, or a similar instrument, with a celestial globe and compass, which will authenticate each observation anticipated by Astronomers, of the situations and bearings of the planets, &c.

On turning the handle of this machine, if we imagine ourselves in the Sun, all will appear regular; but if we consider the motions of the planets, as viewed from the Earth, all those irregularities will be perceived that actually take place, in the observation of them taken from our Earth.

The relative bulks of the balls are to each other, as those of the real Sun and planets are, only not exactly in the same proportion; for the Sun is 877650 times as large as the Earth; Saturn 586 times as large; Jupiter, which is the largest of all the planets, is 1049 times; Mars is five times less than our Earth; Venus very little, if any thing less; and Mercury 27 times less.

To render the celestial phœnomena more generally and better understood, I shall explain each by a diagram, as well as by this instrument; and I hope the repetition will not be found irksome, or be thought unnecessary: I shall explain them as concisely as possible, and am certain it will serve the beneficial purpose of rendering the circumstance of the heavenly bodies more universally intelligible.

I shall begin with the grand vivifying principle, the Sun, and shew you why it appears to pass through all the signs of the Zodiac in one year; and this I shall do, by supposing that we could see the sign which the Sun is in, at the same time that we see him; which, although contrary to fact, yet will serve our purpose of illustration, as we can always know what sign the Sun is in at 12 o'clock each day, by observing what constellation of the Zodiac is on our meridian at 12 o'clock at night, as the Sun must be in the opposite sign to that.

These being understood, you must suppose the exterior circle of *fig.* 2, *plate* 9, to represent the ecliptic, or Sun's apparent path in the Heavens; the circle A B C D to be the orbit of the Earth; and E, F, to be the Earth in two different situations in it; also S the Sun.

When the Earth is at E, the Sun will appear to a spectator on the Earth to be in Taurus. As the Earth moves on easterly in its orbit, and arrives at F, the Sun will be transferred to a more easterly situation as viewed from it, and appear at Leo. And thus, by the annual motion of the Earth, the Sun will be transferred to all the signs of the zodiac, and appear to pass through them all in one year.

This is very evident by the planetarium; for, if we fix the Earth in any situation in her orbit, and turn her forward in it to any other, we shall perceive that the Sun's place will be changing continually in respect to the ecliptic circle of the planetarium or orrery, and that it will enter a new sign every month; circumstances which actually take place in our observations of that body in the heavens, in regard to the constellations, and their locality also accords with the speculum phœnomenorum.

The annual motion of the Earth is the cause of another phœnomenon, which may be understood from *fig.* 2, *plate* 9, or rather the same phœnomenon, only considered in reference to another circumstance produced by it; and that is, we have the opportunity of viewing all the constellations of the zodiac by their being above our horizon, after sun-set, in the course of twelve months.

When our Earth is at E, in *fig.* 2, *plate* 9, that part of it which is turned from the Sun, and appears dark, represents night, and of course the constellations which are opposite to that dark side, are visible to us at night.

When the Earth is at F, the dark part of it being turned towards the constellation Aquarius, that will be on our meridian at 12 o'clock at night. And thus, by the Earth's progressive motion, all the stars of the zodiac are presented to our view in the course of twelve months, a new constellation appearing on our meridian at 12 o'clock at night every month.

This is likewise pleasingly exhibited by the planetarium, which, whilst the Earth is performing its revolution, as it advances shews all the stars of the zodiac that are above the horizon for every night in the year, although it does not shew the other constellations, for which information a celestial globe must be referred to, being adapted for that purpose.

The true motion of the planets, which is according to the order of the signs, is called direct, or in consequentia. The apparent motion is sometimes contrary to the order of the signs, which is called retrograde, or antecedentia.

All the planets sometimes appear to have a retrograde motion, contrary to the order of the signs; or rather, are supposed to have that motion, by being transferred from one place in the ecliptic to another more westerly; whereas, in their direct order, they advance easterly. The apparent retrograde motion of a superior planet, is occasioned by the progressive motion of the Earth, but that of an inferior by its own annual motion; and therefore the places of their retrogression are very different — the superior appearing retrograde when in opposition to the Sun, that is, when our Earth is between them and the Sun; — and the inferior about their inferior conjunction with him, which is, when they are between our Earth and the Sun. The times also of the apparent retrograde motion of superior and inferior planets are different; that of a superior happening oftener to those which move the slowest, but that of an inferior to those which have the swiftest motion; therefore Saturn oftener appears retrograde than Jupiter, and Mercury than Venus; and because a superior planet appears retrograde once in each revolution of our Earth, and an inferior once in each of its own revolutions.

The time of Saturn being retrograde happens at the intervals of one year and thirteen days of our time; of Jupiter, one year and forty-three days; of Mars, two years and fifty-two days; of Venus, one year and two hundred and twenty days; of Mercury, one hundred and fifteen days. Saturn continues in a state of retrogression one hundred and forty days; Jupiter one hundred and twenty days; Mars seventy-three days; Venus forty-two; and Mercury twenty-two, at a mean calculation.

A planet is said to be in conjunction with our Earth, or the Sun, or another planet, when it is in the same point of the ecliptic with either: this is true conjunction. There is also an apparent conjunction. These conjunctions are thus distinguished from each other: —When a line drawn through the centre of the Sun and a planet, passes also through the centre of our Earth, these bodies are really in conjunction; but when this does not happen, although to an observer on the Earth they may appear in conjunction, they are not so in fact. When an inferior planet is between us and the Sun, it is said to be in its inferior conjunction; and when the Sun is between us and a planet, it is said to be in its superior conjunction. The former happens only to the inferior planets; and when they are in it, they cannot be seen by us, but as a dark spot on the Sun's disk.

In referring to the situation of the superior planets, when in a right line with our Earth and the Sun, when they are in the same sign with the Sun, (at which time we cannot see them) the Sun being between them and the Earth, they are said to be in conjunction; and when they are in the opposite sign to that in which the Sun is, they are said to be in opposition—the Sun and planet being on opposite sides of our Earth.

As the planets all revolve round the Sun with different velocities, they must of course change their relative situations to each other.

When an inferior planet is in the same sign or degree of the ecliptic with the Sun, or with the Earth, it is said to be in conjunction with it; and when a superior planet is in the same degree of the ecliptic with the Sun, it is also said to be in conjunction with it; but when it is in the opposite point of the ecliptic from the Sun, that is, in the same sign with our Earth, it is said to be in opposition to the Sun, but in conjunction with the Earth. Any two planets that are in the same sign of the ecliptic are said to be in conjunction with each other.

When the planets are the sixth part of a circle, or 60 degrees, from each other, their situation is denominated sextile. When a quarter of a circle, or 90 degrees, from each other, quartile. When one-third of a circle, or 120 degrees, from each other, trine. When hàlf a circle, or 180 degrees, from each other, they are said to be in opposition.

These situations are called the aspects of the planets; which are expressed in the Ephemeris by characters: Conjunction by ☌, Sextile by ✱, Quartile by ▢, Trine by △, and Opposition by ☍, as exhibited in page 38 of the Ephemeris.

The motions of the planets, and their places in the ecliptic, as seen from the Earth, are called geocentric; their real motions and places in the ecliptic, which are those in which they would be seen if viewed from the Sun, are called heliocentric.

I shall now apply these terms to practice, by which they will be clearly understood, after having explained the different affections and situations of the planets by diagrams.

Let A B C D, *fig*. 3, *plate* 9, represent the ecliptic, which is divided into twelve parts, and on which the twelve signs of the zodiac are delineated. T our Earth in its orbit, and V V Venus in hers, in two opposite points of it; that at V 1, in which she is placed in the same situation or sign with our Earth, is called her inferior conjunction, and that at V 2, represents her superior conjunction. Venus is also represented at E and F, being the places of her greatest elongations or distance from the Sun, as viewed from our Earth, which is forty-eight degrees from it; which are counted on the arch of the ecliptic that measures the angle formed between the straight lines ET FT, or FT ST; one drawn from Venus to the Earth, and the other from the Earth to the Sun; as those two lines form the angle of which the arch G A or A H is the measure—

the Sun and planet both being transferred to the ecliptic circle. This is the difference also between the geocentric place of the Sun and the planet.

Venus never appears to go further from the Sun on each side of him, than to those two situations G and H, or 48°.

Let the circle L m n o represent the orbit of Mercury, and p q the places of his greatest elongations, which are at only 28 degrees distance from the Sun. On considering the situation of these two planets, you will perceive that when Venus is at V 1, and Mercury at m, in their inferior conjunction, those planets cannot appear but as dark spots on the Sun's disk; and that when Venus is at V 2, and Mercury at L, in their superior conjunction, they cannot be seen by us, being in the same sign with the Sun.

When Venus appears west of the Sun, which is from her inferior to her superior conjunction, she rises before the Sun in the morning; and from her superior to her inferior conjunction, or when she is east of the Sun, she sets after him in the evening. She is in each of these situations, alternately, for 290 days.

I shall now explain the foregoing by this instrument, by first placing the Sun, Venus, and our Earth only, on the supporting stem. When I turn the handle, you will perceive that Venus is nearer to the Earth at one time than at another; also that she will appear sometimes nearer to the Sun, as seen from the Earth, at others further from him, and at others in conjunction with that luminary.

When Venus is placed between the Earth and the Sun, she represents her inferior conjunction, and her situation when she appears as a dark spot on the Sun's disk. When on the other side of the Sun, in which she appears in the same sign with it on this instrument, she is said to be in her superior conjunction, at which time she cannot be seen in the heavens. The same may be observed of Mercury.

I shall now take off the balls which represent Venus and the Earth, and place the socket, *fig*. 5, *plate* 10, on the wire which supported Venus; and the socket P, *fig*. 4, *plate* 10, to which a long wire is fixed, on that part which carries the Earth; then, by imagining the brass wire to be the visual ray of a spectator on our Earth, directed towards Venus, which is represented by the white ball at the extremity of the visual ray, and which has the part of the wire connected with it, supported in a groove on the top of *fig*. 5, *plate* 10. This disposition of these parts of the instrument will enable me to shew you, why the planet does not seem to us to recede from the place of the Sun beyond certain limits on each side of him; also the cause of its apparent retrograde motion, which is not actually seen by us, only known by the situation of the planets in respect to the Sun and fixed stars at any time.

Placing now Venus in her superior conjunction; then, as she passes from that to her inferior conjunction, she will appear as supposed to be viewed from our Earth, westerly of the Sun, which is the time of her being a morning star; but as she is moving from 48 degrees on the other side of her inferior conjunction to her superior, she will appear east of the Sun, then is the time of her being an evening star.

I have already informed you that Venus appears retrograde, or to move backward in the ecliptic westerly, once in each of her revolutions in her orbit; I beg you to observe that her time and place of being retrograde, is when she has passed through her superior conjunction eastward, which is her direct and true motion, and arrived to a certain point of the ecliptic near her inferior conjunction, which forms an angle of 48 degrees with the Sun's place as viewed from our Earth, at which place the visual ray from the Earth is a tangent to her orbit; then, if you remark the point of the ecliptic of the orrery at which the planet appears, and observe the motion of the ivory ball after it has arrived at that point on the eastern side of the Sun, you will perceive its motion to be easterly, or direct.

I will remove now the brass ball which represents the Sun, that you may see this more evidently; the cause of this deception is occasioned by not having an intervening object to distinguish the circular motion of the planets.

You now comprehend the cause of the apparent retrogression of the planets; also the cause of its not exceeding certain limits on each side of the Sun; that the former arises from our not being able to judge of the circular motion of those bodies, for want of an intervening object; and the latter, because in two points of the inferior planets orbits, the line of sight, or visual ray of a spectator on our Earth, becomes a tangent to their orbits; that is, touches the extremities of the diameter of their orbits, and thereby con-fines their apparent motions within that diameter.

This is better seen by the shadow of these balls received on a screen, which shews it very clearly; the inconvenience arising from the intervention of the supporting stem being obviated.

The apparent retrogade motion of the planets may also be un-derstood from *fig*. 1, *plate* 11, in which A B C D represents the ecliptic, S the place of the Sun, but that luminary invisible; T the Earth in its orbit, and V G H Venus in hers; also, at V, the place of her superior conjunction. Let us suppose her to be moving from that situation easterly, till she arrives at H, near her inferior con-junction; then will her place in the ecliptic appear at E, as seen by a spectator on the Earth.

Now, suppose the line 1 2 to be the visual ray of an observer on the Earth, this ray being a tangent to the orbit of Venus, that is, touching one of the extremities of its diameter, as she is com-pleting the other part of her revolution from H to G, till she ar-rives at G, where the ray again becomes a tangent to her orbit; she will appear to move amongst the fixed stars backward, or contrary to the order of the signs, from east to west, because there is no intervening object by which to judge of her circular motion.

When an inferior planet is in the superior part of its orbit, its motion appears true and direct, from west to east; when in the inferior, false and retrograde, from east to west.

In referring to this diagram, I have taken no notice of the motion of our Earth in its orbit, as it would only confuse the mind by multiplying ideas, and as the relative situations of the two bodies are as well understood without it.

All that we have observed of Venus and our Earth, is equally applicable to Mercury; allowing that, on account of his distance from the Sun being less than that of Venus, his change of motion must happen oftener, as it depends also on his arriving at the two extremities of the diameter of his orbit, the same as that of Venus does; and therefore, on account of his orbit being less than that of Venus, and his motion in it quicker, the circumstance alluded to must occur oftener to him.

At the points of a planet's orbit, in which it appears to change the direction of its motion, it seems to remain some time stationary, because a part of the curve is so near the point which the ray touches, that it cannot be distinguished from it.

One thing remains to be observed of *fig.* 3, *plate* 9, in which I have considered the Earth as at rest, and Venus in motion, which serves to convince us of the absurdity of supposing the Earth to be actually at rest in the centre of our system; because, if it were so, the planet would appear to pass through all the signs of the Zodiac in its revolution; or supposing it to be at rest at all, even in its situation represented *fig.* 1, then the planets would always be seen in the same signs of the Zodiac, as between F and E; whereas they are seen in all, occasionally, according to the motion of our Earth compared with their motions.

This instrument is particularly well calculated for exhibiting the phases of the planets of our system, which shine only by reflecting the light of the Sun; and therefore, by fixing them for their situation at any time, in respect to the Sun and our Earth, and turning them on in their orbits, their different phases, as seen from the Earth, will be exhibited.

We will, if you please, first, make observation of the phases of Venus; as, on account of her near situation to our Earth and the Sun, they are more distinguishable than any other. If we place Venus at either her superior or inferior conjunction with the Sun, and turn her on in her orbit, as the Earth and planet revolve round according to their respective velocities, the ivory ball will exhibit the phases of Venus as seen from the Earth. Thus, if we wish to know what phase that planet will exhibit for any evening on which she will be visible, we must place the Earth and Venus in their respective places on the planetarium for that evening, (by the information of an Ephemeris,) and after fixing both balls with the white sides towards the Sun, we shall see the phase of Venus as it will be actually perceived by us through a telescope, or very nearly so, allowing something for the inclination of her orbit, or place of the nodes, this instrument not exhibiting those deviations.

The phases of Venus may also be understood from *fig.* 3, *plate 9.* V 1, is Venus with her dark side turned towards the Earth, which renders her illuminated side invisible to us. When seen, she appears as a dark spot on the Sun's disk, which is called her transit; this does not happen every time she is in her inferior conjunction, because of her orbit being inclined to the plane of the ecliptic, or central solar ray, and therefore she does not always pass in a direct line between the Sun and us; and when she does not nearly do so, on account of her distance from the Sun and us, she does not appear to pass over his disk. When she is in her superior conjunction, as at V 2, *fig.* 3, *plate 9,* she has the whole of her illuminated surface turned towards the Earth; but on account of her nearness to the Sun at that time, if not in a direct line with the central solar ray, she is not seen by us in either case; in one, because of her affinity of situation to that luminary, and in the other, on account of the solar ray intercepting our view of her.

When Venus is near to her superior conjunction, she appears nearly round to us, but small, on account of her distance from the Earth. When she is east of the Sun, which happens immediately after she has passed through her superior conjunction, she appears larger every day, because she is approaching nearer to the Earth each day.

The superior as well as the inferior planets, are sometimes nearer to the Earth than at others, and consequently appear larger at those times at which they approach nearest to it. They have also the appearance of being stationary and retrograde, as well as the others; yet, by describing an orbit much larger than ours, and exterior to it, they constantly turn the greater part of their illuminated side towards us, and therefore appear almost like the Moon at full.

The superior planets are retrograde when seen in opposition to the Sun, that is, when our Earth is interposed between them and that luminary; and they are direct when in conjunction with the Sun, that is, when the Sun is between us and them.

Let ABCD, *fig.* 2, *plate* 11, represent the orbit of Mars, (as it is immaterial which of the superior planets is nominated to illustrate their general motions and appearances,) EFGH the orbit of our Earth, and S the Sun. Then, when Mars is at I in his orbit, and the Earth at K in hers, the Sun being between the planet and the Earth, Mars will be in conjunction with it. But when Mars is at L, and the Earth at K, the Sun and planet being on contrary sides of the Earth, the planet is said to be in opposition to the Sun.

In explaining the retrograde motion of a superior planet, by *fig.* 3, *plate* 11, we must consider ABCD the ecliptic, NMRS the orbit of Mars, O P Q T the orbit of our Earth, and S the Sun, in the centre. As the superior planet is much longer in performing its annual revolution than the Earth, and as its apparent retrograde motion depends on the diameter of the orbit of our Earth, it will be more natural to suppose Mars to be at rest in this illustration, and the Earth in motion; therefore, let us suppose Mars to be stationary, as at N, in opposition to the Sun, and let the Earth be first considered, as at O, when it is exactly in a direct line between the Sun and Mars. In this situation of Mars and our Earth, the former will appear to us to be in the opposite sign to that in which the Sun appears; but if the Earth be supposed, as at O, in its orbit, the visual ray to a spectator on it, as directed to Mars, will be transferred to A in the ecliptic. If we consider the Earth at either P or Q, a line drawn from it through the planet to the ecliptic, forms a tangent to the Earth's orbit. Whilst the Earth is moving from Q to P, through 123456, and its conjunction with the Sun at T, its motion is direct, or from west to east, in the ecliptic, as may be seen by the lines drawn through that part of the Earth's orbit, through Mars to the ecliptic. When he arrives at P, the visual ray is directed to K in the ecliptic; and in moving through 654321, and its opposition N, it will appear to describe the arch of the ecliptic R I H G A F E D C, and to move from east to west, or retrograde, through the same portion, of the ecliptic. When the planet is either at P or Q, it will appear stationary for a few days.

I hope I have now conveyed to you a clear idea of the real and apparent motions of the planets. One thing yet remains to be illustrated by the planetarium, which is, the absurdity of the Ptolemaic system, and the truth of the Copernican. I will place the ball which represents our Earth on the centre stem of this in-the other planets in their proper situations. This disposition of the heavenly bodies is conformable to the Ptolemaic System of the Universe; and on putting the bodies (thus arranged) in motion, we shall perceive that were our Earth so situated, the motions of all the planets would appear regular and direct to us, and that every other phœnomenon observed of them, as their periods, phases, &c. would be exactly contrary to what is actually perceived of them. Venus and Mercury might also be seen in opposition to the Sun, or in the opposite sign to him; and the superior planets when in the same sign, or in conjunction with the Sun.

My desire of introducing to my dear pupils these studies, arises from a conviction of their utility, inasmuch as they elevate the mind by the communication of ideas naturally tending to refine and purify the imagination; leading it to reject frivolous and low pursuits, and to delight only in such things as exalt and perfect human nature — by which means they will avoid engaging in the follies of those thoughtless beings who waste their days in baneful pleasures, which finally terminate in mortification and disgust.

That active principle, the imagination, must be employed on some object or subject; it is therefore the duty of those who undertake to form the minds of youth, to guard them against those which are improper: this duty can no way be so efficaciously performed as by introducing them to those which are not only proper, but improving and interesting; and that this is essential, the many instances we meet with of prostituted abilities, both in society and literature, clearly demonstrate.

How much then have those to answer for, who, possessing superior sense, not only neglect the proper use of it, but employ it in such a manner as to deprave human nature.—Never, my dear young friends, suffer yourselves to read an author, be his abilities ever so great, who wishes to instruct you in what is contrary to virtue, by painting vice in false colours; for such is the infirmity of our nature, that it is not safe to be familiar with books, or conversation, which have the least immoral tendency, because the effects of habit are so powerful, that what at first may disgust us, (by reason of the native dignity of virtue which is inherent in every uncorrupted breast) may, by the mind becoming accustomed to it, and losing that delicacy natural to it, cease to disgust us;—and when the finer impressions are thus injured by the rough and coarse ideas impressed on the mind, they are not only deprived of their energy, but too often no vestige remains.

How important does it appear then, to cherish all ideas that elevate the mind, and lead to virtue! It is the only sure barrier against the encroachments of folly and depravity, and is also the most graceful ornament of our nature; for if we wish to please by an engaging exterior, there is no surer method to obtain that advantage, than by furnishing our minds with ideas which are beautiful and harmonious—and dignity of character will always result from elevation of sentiment.

We can judge, by the performance of any artist, whether he has cultivated a good or bad taste; and, if we perceive the former, we not only admire the subject, but the man; but if the latter, although equally well executed, we are not pleased with the performance, nor can we esteem the artist; as we suppose, in both cases, that the character of the piece is descriptive of the genius or disposition of the person who produces the effect.

Thus will you become esteemed and admired, in proportion as you cultivate a taste for the sublime and beautiful, which you can only attain by an innate sensibility to each, and which, when possessed of, will never fail to produce the sweet, the charming emanations of inherent excellency.

END OF THE EIGHTH LECTURE.

Plate XII.

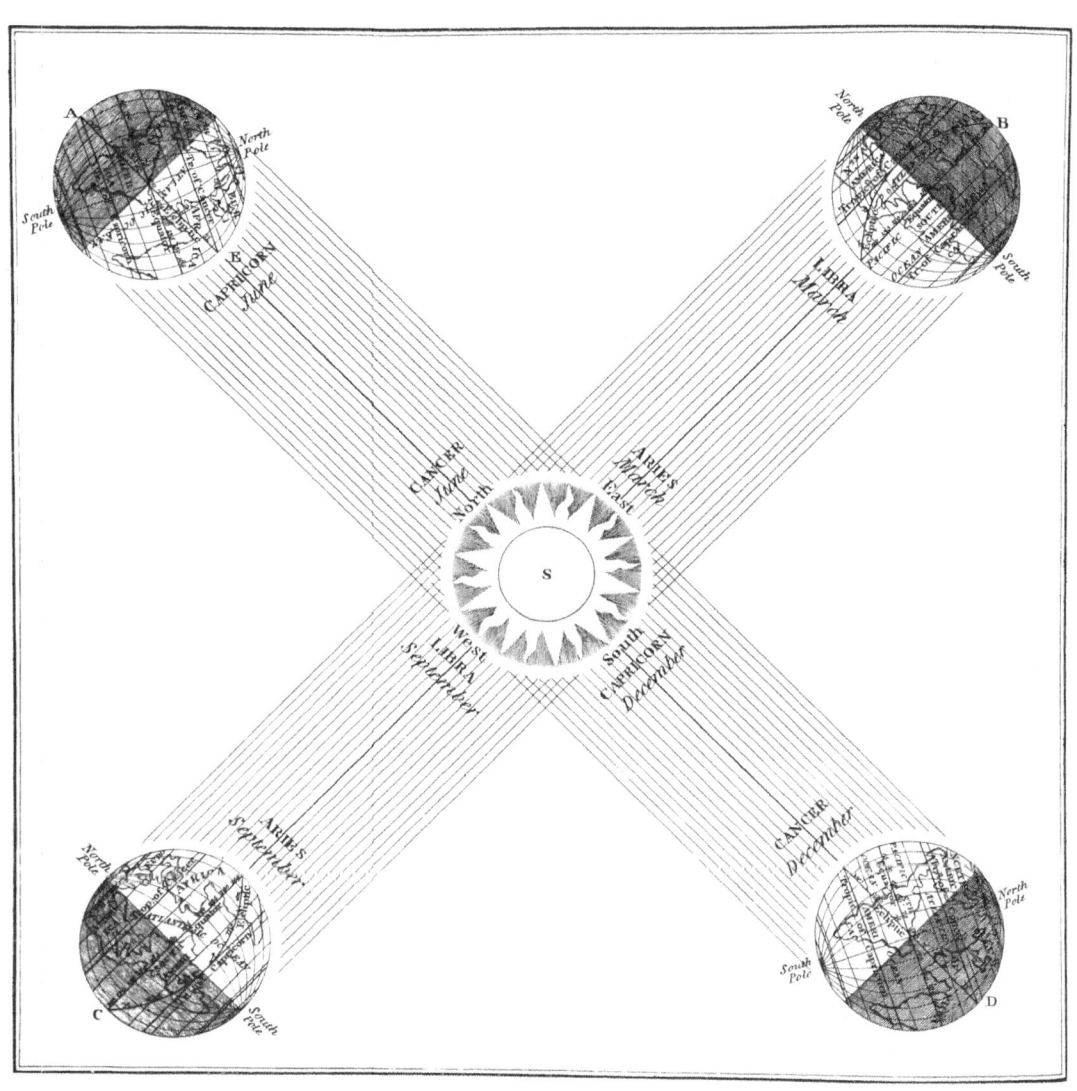

T. Conder Sculp.ᵗ

Plate XIII.

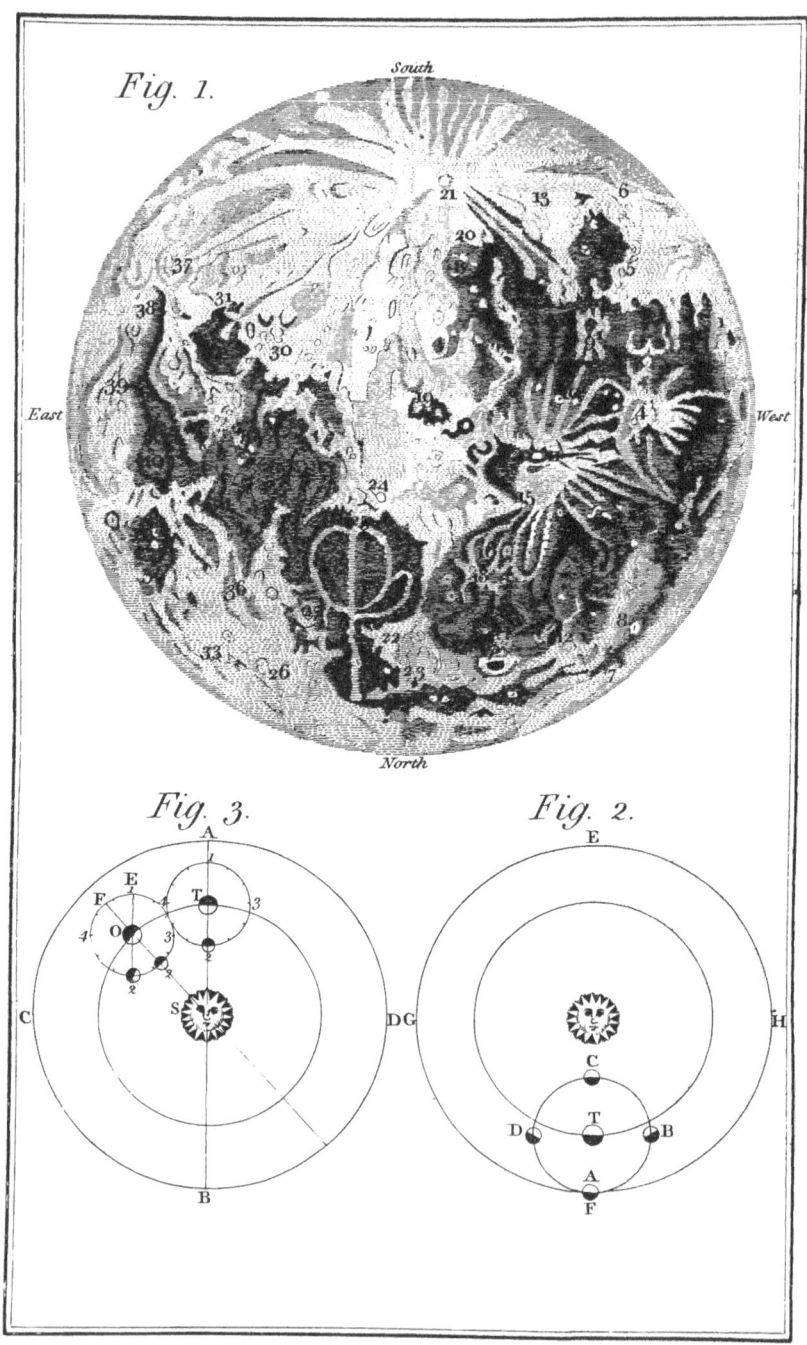

LECTURE IX.

OF THE OBLIQUITY OF THE AXIS OF THE EARTH TO THE PLANE OF ITS ORBIT.—THE NUTATION OF THE EARTH'S AXIS AC-COUNTED FOR.—OF MR. NORWOOD'S METHOD OF ASCERTAIN-ING THE NUMBER OF MILES CONTAINED IN A DEGREE OF THE EARTH; AND THE IMPROVEMENTS MADE SINCE HIS TIME IN THIS CALCULATION, BY A KNOWLEDGE OF THE TRUE FORM OF THE EARTH ON THE PRINCIPLES OF GRAVITY.—OF SOLAR AND SIDEREAL TIME, AND THE INEQUALITIES OF THE SOLAR DAYS AMONGST THEMSELVES.—OF THE TROPICAL OR CIVIL YEAR.—OF THE VICISSITUDE OF THE SEASONS, AND THE DIF-FERENT LENGTHS OF DAY AND NIGHT THROUGHOUT THE YEAR.—OF THE MOTIONS OF THE MOON.—HER DIFFERENT PHASES.—THE INCLINATION OF HER ORBIT TO THE ECLIPTIC.—THE GREAT VARIETY IN HER AFFECTIONS.—OF THE SYNO-DICAL AND PERIODICAL MONTH.

————————— Not content
With every food of life to nourish man,
By kind illusions of the wond'ring sense
Thou mak'st all nature beauty to his eye,
Or music to his ear.——— AKENSIDE.

BY removing the planets, and placing *fig.* 2, *plate* 10, on the plate of the orrery, it forms a tellurium, which is constructed to exhibit the phœnomena occasioned by the diurnal and annual motions of the Earth, and the obliquity of its axis to the ecliptic, with every circumstance in which the Sun's influence on it is individually considered.

The globe turns on an axis, which is inclined to the plane of its orbit, as that of our Earth is, in an angle of 66¼ degrees, which is the distance of the north and south extremities of it from the solstitial points, those which are the Sun's limits on each side of the equator, as will be perceived (when the Earth is turned on its orbit) by observation of the solar ray G.

RS represents one half of the circle of longitude, as I have already mentioned, and RST the moveable meridian; the former should always lay over the equator of this globe, as the longitude of places on it are referred to the equator; and the latter will then shew the latitude of all places on this globe, which is their distance from the equator. When we want to know the longitude of places, we set the first degree of the circle RS in a line with London, that being the place from which longitude is counted in England, and then all the degrees of longitude will answer to their proper places on this globe. If we want to know, at the same time, the latitude of all the places on the globe, we must observe the horizontal lines from each degree of latitude, marked on the semicircle RTS, and the places will each correspond with their respective degrees of latitude:—or we may, after finding the longitude, move the globe round, so that the general meridian may lie over the different meridians of places on the Earth.

The dial under the globe shews sidereal time, that in which the Earth revolves on its axis, which is always performed in 23 hours, 56 minutes, and 4 seconds. The dial under the Sun shews solar time, or that which elapses between the Sun transitting our meridian one day, and his return to it the next day;—the effects of these circumstances I shall explain in their proper place.

When we fix the Earth on any part of the ecliptic plate of the orrery, the Sun's place on it will be transferred to the opposite sign.

As it is by the Sun's place that we usually set astronomical instruments, in order to observe terrestrial and celestial phœnomena, in using this instrument, I look for the Sun's place, and letting the index K point to it, set the Earth to the opposite sign. On turning the Earth on in its annual orbit, the index K moving along at the same time, transfers the situation of the Sun to every point of the ecliptic circle, and keeps the Earth and Sun in their respective situations during the Earth's intire revolution in its orbit.

The terminator H I, which divides the enlightened from the dark hemisphere, together with the solar dial, serve to shew the length of the day and night of every place on the globe, for any day in the year, by setting the globe to the situation of those places for any given time, in respect to the solar ray; which is easily done, by taking the situation of the Earth at any time as a standard.

Previous to applying these observations, I have a few things to acquaint you with respecting the Earth, which I omitted introducing before, thinking it might afford me the opportunity of a necessary revise of a subject much treated of in my preliminary Lecture. I then informed you of a mode used by the ancients to measure the size of our Earth, and also pointed out to you the imperfection of that mode, originating in their ignorance of gravity and geometry.

We will now examine the accuracy of the modern mode of estimation, resulting from the knowledge of gravity and mathematics.

Many methods have been devised for determining this fact of the size of the Earth; but I shall select the one pursued by Mr. Norwood, and practised in 1635, as it was more successful than any that had been previously adopted, and is easily understood, being divested of those niceties of calculation since entered into, which, although improvements on the accuracy of his, yet in describing the mode he pursued, we are less perplexed, and equally gratified, the principles of estimation being the same in all. Before following Norwood in his investigation of this subject, I will inform you, that by the inferences deduced from his mode, and from the improvement of others since his time, in calculation, who have allowed for the oblate figure of the Earth, and have used instruments of greater nicety of adjustment, by which the quantity of a degree has been ascertained with more precession; it appears, that the Earth is about 25,000 miles in circumference, 7,978 in diameter at the equator, and 7,938 at the poles; also that its surface contains 198,950,000 square miles.

Mr. Norwood's manner of estimating the length of a degree on the Earth's surface, and its circumference, was this:—He took the Sun's altitude when on the meridian at London, on the day of the summer solstice, and then made the same observation of the Sun's altitude on the meridian at York, on the year following. He thus discovered by the difference of its altitude at the two places, that their difference of latitude was 2° 28′. He likewise measured the distance between the two cities, or their difference of latitude in miles. This must have been very tedious, as at every curviture of the road, he must have made a drawing, and have calculated the sides of all the angles and triangles, in order to reduce the whole into an uniform arch of a circle, allowing also for the ascents and declivities in the road. These two places lying nearly under the same meridian, or having nearly the same longitude, when the above-mentioned circumstances had been allowed for, also the latitude of the two places, the angle of the arch of a circle formed on the scale corresponding to the great circle of the Heavens, the relative portion of a whole circle on the scale was clearly seen; and of course, by the quantity it contained in English miles within that arch, compared with the degrees of a circle, it was discovered how many English miles a degree contained.

When this observation was taken, the figure of the Earth was not considered otherwise than as a perfect sphere. But since his time, all impediments to accurate computation have been overcome by Picard, Cassini, and other measurers, who have taken observations at many different degrees of latitude between the equator and the poles. Thus we find that perseverance will surmount all difficulties arising from natural effects.

This globe, you perceive, by the degrees on the brass semi-circle, is inclined to the centre of the Sun or plane of its orbit, in an angle of almost 66 ½ degrees, which is nearly its constant inclination to it; although it varies a little, but not more than 19 seconds of a degree in nine years; and the motion which causes this variation is called the nutation of the Earth's axis.

This motion of the axis of our Earth is produced by the attraction of the Sun and Moon upon the protuberant matter about the equator. Although this motion is dependant, both on the attraction of the Sun and Moon, yet it is effected chiefly by that of the Moon, on account of her nearness to the Earth, and of her deviating farther from the equator; by which her action is as it were perpetual. These two attractions, being unequal to each other, cause this motion of the axis of our Earth to be performed in a conical shape, the extremity of the axis forming a small elliptic revolution round the centre of their mutual attractions, which is completed in 18 years and 7 months, being nearly the same with the period of the Moon's nodes, which I shall speak of hereafter; upon the motions of which the quantity of the nutation of the Earth's axis each year principally depends. The period of its intire revolution, or time of the axis of our Earth forming exactly an angle of 66 degrees and a half with the ecliptic, or central solar ray, has been fixed by observation, taken at different times of the declination of the fixed stars, as well as by calculation on Sir Isaac Newton's principles of attraction. As the orbit of the Earth is an ellipse, it is certain that it must be nearer to the Sun in some parts of it than in others; also, that as its axis is inclined to the centre of the Sun, when the Earth is situated in those parts of its orbit where the equator is in a straight line with the Sun, which happens in only two points of it; the action of the Sun cannot affect the axis of the Earth so as to incline it more or less from its perpendicular situation, as it acts equally on the two extremities of the axis; but when it is in any other part of its orbit, the Sun acts obliquely, and unequally, on the protuberant matter about the equator, and of course the inclination of the axis of the Earth to the centre of the Sun must be altered; and the effects of this variation, being the same on both sides of the equator, it occasions the extremities of the axis to form an ellipse. The nutation of the Earth's axis has been computed by Astronomers for every time of the year, by comparing the attractions of the Sun and Moon together, and is set down in the Ephemeris; at least, the different angles formed between the equator and ecliptic is noted for different times of the year, which depends on this circumstance, united with the precession of the equinoxes.

When we reflect on the variety of circumstances which must be allowed for, by all Astronomers, in their observations of the phœnomena of the heavenly bodies, we are astonished at the success and patience with which they prosecute their researches. They must certainly be endowed with a most extraordinary spirit of investigation, and firmness of intellect, to surmount all obstacles, without regard to the length of time passed in these amazing exertions of body and mind. Those men deserve our admiration,

> Whose curious thoughts with active freedom soar,
> And trace the wonders of creating power. Mrs. E. CARTER.

I will now proceed to explain the phœnomena this machine is constructed for exhibiting, both by it and by diagrams.

The Earth's motion on its axis, as I have before observed, is always equal, and is performed in 23 hours, 56 minutes, 4 seconds, and therefore the sidereal days are always precisely of the same length; that is, any star perceived on our meridian at 12 o'clock one night, will appear in the same situation at 56 minutes 4 seconds past 11 the next night, by a well regulated clock. But as the Earth has a progressive motion also, by which it advances almost a degree eastward in its orbit, in the time it turns once on its axis, the Sun does not perform his apparent revolution, or his reappearance on our meridian is not effected, in exactly the time a given star performs its apparent revolution in.

The solar days are not all equal among themselves, on account of the obliquity of the equator to the ecliptic, which causes the Sun to be sometimes longer than others in the time of two of its conjunctions with a particular meridian; therefore the solar days are sometimes more and sometimes less than 24 hours, never being exactly that length but on four days in the whole year, as at all other times the central solar ray is inclined to the equatorial motion; but on those four days it is parallel with it, and therefore the Sun will transit the meridian exactly in the 24 hours.

The Earth turns 366 times on its axis whilst the Sun performs 365 of his apparent revolutions, or the Earth the 365 parts of its orbit; because as the Earth advances almost a degree, or the 365th part of its orbit, eastward, whilst it turns once on its axis, it requires as much more than one turn on its axis to complete the natural or solar day as it has advanced in its orbit in that time—therefore we have one more sidereal than solar day in the year; consequently the sidereal days, or the times that our Earth revolves on its axis, in performing its annual revolution, are 366 in the year. The inequality of the solar days amongst themselves is occasioned by the obliquity of the orbit of our Earth to the plane of the equator, which occasions the Earth's motion in it (and on which they depend) to be sometimes swifter and sometimes slower than the mean rate of 24 hours.

The Ephemeris contains tables of the equatation of time, which is the difference between mean time, or that shewn by a well regulated clock and solar time, for all other times except the four days already mentioned.

The reason of the motion of the Earth on its axis bringing the same degree of the equator to our meridian at the same time each evening, is because the equator is perpendicular to the axis of its motion; whereas the ecliptic, or Sun's apparent path, being inclined to that motion, the equal motion of the Earth on its axis, which brings equal portions of the equator to the meridian in equal times, must bring unequal portions of the ecliptic, and the difference is in proportion to the obliquity of the ecliptic to the equator.

The popular mode of illustrating this position, is by placing patches, at equal distances from each other, on the equator and ecliptic of a globe, beginning both from the first point of Aries, and continuing them through one half of both those circles, and at the same distances throughout. Then supposing two Suns to be starting together from Aries, one in the ecliptic, and the other in the equator; on turning the globe, all the patches from Aries to Cancer, on the ecliptic, will arrive at the meridian before those on the equator; which shews that the Sun, during that time, is before the clock. When Cancer is at the meridian, the patches on the equator and ecliptic answer to each other, and arrive at the meridian at the same instant; therefore the Sun and clock will be together on that day.

From Cancer to Libra the patches on the ecliptic will arrive later at the meridian than their corresponding ones on the equator; which shews, that whilst the Sun is in the intervening signs between Libra and Cancer, he will appear on our meridian later each day than 12 o'clock. This shews one difference of the solar days amongst themselves, or the equatation of time. By taking off the patches and fixing the index to any hour, and only observing it at the different periods and stations, you will see the variation between solar and sidereal time; by remarking any point of the equator at the meridian and the degree of the ecliptic at it, at the same time, and turning the globe round, until those two points again arrive at the meridian, you will perceive that they return to the same situation exactly in 24 hours. But as the Earth advances almost a degree easterly in its orbit in that time, the place of the Sun must be transferred one degree east of the Earth's place on the equator; and when that degree of the ecliptic comes to the meridian, you will perceive that the Sun's place has taken four minutes more in arriving at the meridan than the Earth's place. This illustrates the difference between solar and sidereal time, although this globe always brings the same degree of the equator to the meridan in 24 hours instead of only 23 hours 56 minutes; therefore all that must be observed as accurate in this problem, is the difference of time, not the absolute length of time, in which the Earth revolves on its axis.

When the Sun is at the first of Aries, the first of Cancer, the first of Libra, and the first of Capricorn, being parallel with the Earth's equable motion, the same degrees of the ecliptic and equator will be on the meridian at the same time.

I have taken the liberty of altering the illustration of these problems a little from the original, and also of introducing some auxiliary circumstances, for the purpose of rendering it more perspicuous.

When the Earth is advancing towards the Sun, its motion is accelerated; but when it is receding from that luminary, its motion is retarded. The circumstances of the accelerated and retarded motion of the Earth happen alternately.—In one half of its orbit it is the former, and in the other half the latter; therefore the solar and equal days do not agree exactly in the time of the day. except on four days of the year; that is, when the Earth is in those two points of its orbit at which it is at its greatest and least distance from the Sun — the first denominated apogee or aphelion, and the latter the perigee or perihelion, and when in the two medium points between these two. The difference between solar and mean or equal time depends on the unequal motion of the Earth in its orbit, compared with the equable motion of the Earth on its axis, whereby they do not perfectly agree together in the time of the day but four times in the year, namely, on the 14th of April, the 15th of June, the 31st of August, and the 21st of December; and the inequality of the solar days amongst themselves, depends on the different obliquity of the plane of the Earth's orbit to the ecliptic in different parts of it, whereby it is not in the same plane, or without some obliquity, except four days in the year, on which four days the Sun and the same degree of the equator transit the meridian at the same instant. These four days do not happen always at the same times, on account of the intercalary day introduced every fourth year; but the real times of their being at these four points of the ecliptic each year, are shewn by the Ephemeris.

The tellurium pleasingly and conveniently exhibits the difference between solar and sidereal time.

I will fix the Earth on the first degree of Capricorn, being that point of the ecliptic opposite to the Sun's place, for (we will suppose) the 21st of June, fixing the time of his being in apogee or aphelion, or at his greatest distance from the Earth, to be on that day in working this problem, although it does not always happen on the same day of the year, for reasons previously given. This situation of the Earth and the Sun answers in this case to the longest day in northern latitudes, and I fix on the 21st of June, as its deviation from it is very trifling. I will now rectify the globe by means of the arm K L, *fig.* 2, placing the northern polar circle intirely on the illuminated side of the terminator, and I will turn the handle of the machine till the index under the globe points to 12 o'clock of idereal time; I then place the index on the sun-dial to 12 o'clock of solar time. As I turn the Earth on in its orbit easterly, you will perceive for a few days the dials will nearly correspond in time, on account of the little difference of their respective motions; although the index which shews sidereal or rather mean time, will point to 12 each day, a little before the solar one. This difference of motion will be continually increasing, so that when the Earth has performed half its annual revolution, and the Sun appears at Capricorn, the sidereal dial will have gained 12 hours on the solar one. This situation of the Earth represents its perihelion, or nearest distance from the Sun, as represented at B, *fig.* 1, *plate* 9; which is also the time of the shortest day in northern latitudes.

Whilst the Sun is compleating the other half of his apparent annual revolution, from Capricorn to Cancer, the sidereal dial will continue to gain upon the solar one; so that when the Sun is arrived at Cancer, the sidereal index will have advanced through the whole 24 hours, gaining a whole day in the intire revolution of the Earth in its orbit.

The situation of the Sun when at Cancer is that of his greatest distance from the Earth, or aphelion, as represented at A, *fig.* 1, *plate* 9. This is the time of the longest day in northern latitudes.

I hope I have explained this in a rational way, so as to make these circumstances better understood than they generally are. One thing remains to be observed of the effects of this instrument, as it exactly corresponds with the actual cause of the difference in solar and equal time, which is, that the reason of the dial's shewing the two times, is because the sidereal advances along with the Earth in its orbit, at the same time that it corresponds with its diurnal motion: the Earth's axis being made to alter its inclination to the central solar ray, as it actually does in the Heavens, (although the

Earth nearly preserves its parallelism, in regard to the plane of its orbit, which is always about $66\frac{1}{4}$ degrees,) its inclination to the central solar ray is variable ; and it is this which causes the Sun to appear to move north and south on our meridian, which the central solar ray of this instrument shews—and of course, when he is at the highest point on it, he must remain the longest time with us, by reason of the great angle he makes with the horizon ; and when at the lowest point on it, he must remain the shortest time above the horizon ; and at all the intermediate degrees, intermediate lengths of time.

On this instrument, the teeth in which the wheel works, which gives the Earth its progressive motion, are disposed in the manner, and according to the number of degrees, into which all circles are divided, namely 360 ; therefore the Earth and dial advance nearly a degree of that circle in each revolution of the Earth on its axis ; so that at the end of its annual revolution, it has turned 366 times on its axis in performing it, thus according with sidereal time. But our year is not reckoned according to either of those times, but according to what is called tropical, being computed by the period which elapses between the Sun being on the solstitial point of Capricorn one year, which is in December, and his return to it the next, which is found to be performed in 365 days, 5 hours, 48 minutes, and 54 seconds. This is called the civil year, and is that observed by all nations. But this civil year is sometimes called a common year, and sometimes a leap year.

In a common civil year, the odd hours and minutes are left out of the account, and it consists of 365 days, containing only seven months of 31 days, four of 30 days, and one of 28 days. But then, to make up for the odd hours and minutes omitted in the common years, every fourth year is made to contain 366 days, which brings the time nearly to answer to the course of the Sun. Its deviation from that, I have informed you of before, and also how it is allowed for.

Having explained the causes of the difference of sidereal and solar time, and illustrated them by this instrument, I will now exhibit the causes of the vicissitudes of the seasons, as also the different lengths of the days and nights throughout the year.

The Earth is now again on the first degree of Capricorn, and the annual index pointing to the first degree of Cancer, which is the situation of the Earth for the longest day in northern latitudes; the whole of the northern polar circle being on the illuminated side of the terminator.

London being the capital city of England, I will make the meridian of that place to correspond with the central solar ray, placing the brass semicircle over it; and then the first degree of longitude on the brass circle must be supposed to be at the meridian of London, the celestial and terrestial longitude not agreeing; for London is removed many degrees from the first degree of celestial longitude, which is always at the intersection of the equatorial and ecliptic.

The meridian of London being thus placed opposite the central solar ray, and the solar index pointing to 12 o'clock, denoting the time of the Sun being on the meridian when he is in the first degree of Cancer, we shall perceive, by observing the time on the sun-dial of the spot which represents London on this globe, transitting the terminator, the time of sun-setting at London, on that day in which he is in the first degree of Cancer. Then noting the time of London being on the dark side of the terminator, we shall know the length of the night; and on noticing the time of its appearing on the light side of the terminator, we shall perceive the time of sun-rising; and lastly, comparing our last observation with the time London again arrives on the dark side of the terminator, we shall know the length of the day at that period. At the latitude of London, and all places on that parallel, this will be the longest day in the year, and the shortest night. In like manner, the length of the days and nights at any place on our globe may be known, by observing its latitude, which this moveable brass meridian will shew, by reference to the parallels of latitude, and meridians or longitude; for by fixing this globe for any one place, and for any one time in the year, we may know the situation and circumstances of all other places, without further trouble of rectifying the globe.

In pursuing our present investigation, we will fix upon London to make particular observation of the length of day and night, and of the seasons. When the instrument is adjusted to the situation of London for the 21st of June, the Sun's place appearing in Cancer, all the parallels to the equator on this globe will be cut by the terminator into two unequal parts, excepting those contained within the northern and southern polar circles.—When the Sun is in the first degree of Cancer, the whole of the northern polar region is on the illuminated side of the terminator, and therefore we know that the inhabitants of that region must at that time, when the Sun is in the first degree of Cancer, have perpetual day, never losing sight of the Sun;—whereas those of the southern polar region must have perpetual night. But you will observe, as the Earth performs its annual revolution, these regions have, in some parts of them, variety in the length of their respective days and nights, and that it is only at the extremity of the poles that they have a continuance of such days and nights as I have been describing; also that at the northern pole they will have six months day, whilst the Sun is passing from Aries to Libra, and at the southern pole six months night; but that as the Sun passes from Libra to Aries, these circumstances will be reversed to each.

The Sun, when in the beginning of Cancer, is perpendicular to the tropic of Cancer, and therefore in the zenith of all the inhabitants of that parallel, of course his rays must fall perpendicularly and abundantly on them.

If the axis of the Earth were not inclined to the plane of its orbit, we should not have the pleasing and useful vicissitude of seasons that are now afforded us.

All the planets of our system, of the diurnal motion of which we are assured, have their axes inclined to the plane of their orbits, but that of Jupiter being very little, his days and nights must be nearly equal all the year through; therefore he cannot experience much variety of seasons. As the Earth advances in its orbit, by preserving its parallelism, the inclination of its axis, and the situation of places on it, in respect to the Sun, are continually changing, as you will perceive whilst I am turning the globe on in the ecliptic of this instrument; for as it advances easterly from Capricorn to Aries, the northern polar circle will recede further and further from the Sun, getting more and more on the dark side of the terminator; so that when the annual index points to the first of Libra, which is the time of the autumnal equinox, the terminator will divide the northern polar circles into two equal parts, and also all the parallels to the equator; consequently the days and nights must be equal to each other all over the globe. In this situation of the globe, you will perceive that the Sun acts equally on every part of the protuberant matter about the equator, and therefore that the axis can suffer no nutation. The Sun is now perpendicular to all the inhabitants of the equator.

On turning the Earth on in its orbit to Cancer, the days are found to be perpetually decreasing to the inhabitants of London, and to all the inhabitants of northern latitudes. When the Earth arrives at Cancer, and the annual index points to the first of Capricorn, the day will be the shortest at London of the whole year, and of course the night the longest. The length of both will be

D d

shewn by the sun-dial; also the whole of the northern polar circle will now be on the dark side of the terminator, and the whole of the southern on the illuminated side of it. The central solar ray, when the Sun is in the first of Capricorn, points to the tropic of Capricorn, and the days at London are then shorter than the nights.

By turning the Earth on till the Sun is in Aries, we again see how all the parallels to the equator are equally divided by the terminator, this being the time of the vernal equinox; so that the days and nights are again equal to each other all over the globe, and the Sun rises and sets at six o'clock at every place on it. From this time, till the Sun is again in Cancer, the days at London will be continually increasing in length.—Thus you have had a full view of all the phœnomena of our Earth, dependant on the inclination of its axis. For the information of those who may not have the opportunity of exemplifying these circumstances, by this or similar instruments, I will endeavour to explain the vicissitude of the seasons by a diagram.

Let *plate* 12 represent the situation of the Earth and the Sun, at the four quarters of the year.—In this figure, S is the Sun, A the situation of the Earth in respect to the Sun in June; in which situation of it, the Sun's place is transferred to Cancer, and the whole of the northern polar circle of our Earth is turned towards the Sun, and therefore those regions must enjoy perpetual day. We also observe, that the central solar ray (which is represented by the broad line E F,) points exactly to the tropic of Cancer; and therefore, that the inhabitants of that tropic receive the direct influence of the Sun's rays at this time. The line which divides the dark from the enlightened part of the globe, now cuts all the parallels to the equator, excepting the polar circles, into two unequal parts, and therefore the days and nights are unequal to each other, in all those parts not included within the polar circles.

When the Earth has arrived at C, that part of its orbit in which it is at Aries, and the Sun is in Libra, the time of the autumnal equinox, the equator and the parallels to it being cut into two equal parts by the terminating line, the days and nights are of an equal length all over the globe. In this situation of our Earth, the central solar ray points directly to the equator, and is therefore perpendicular to the inhabitants of that latitude.

As the Earth moves on in its orbit, the northern polar circle will continually be receding more and more from the Sun; and when the Earth has arrived at D, or at Cancer, and the Sun appears in Capricorn, the northern polar circle will be wholly on the dark side of the Earth, and the southern polar circle on the enlightened side. The circumstances of those regions will be exactly the same, whilst the Earth is going from hence to Capricorn, as was observed of the northern polar regions, whilst the Earth was passing from Capricorn to Cancer; and those of all the places in southern latitudes will, in the mean time, be the same with those which were observed of all the places in northern latitudes, in the foregoing half of the Earth's revolution; which circumstances it is unnecessary to repeat, as the application to them has been fully made in the illustration of this subject, by means of the tellurium.

The next object which offers itself to our consideration, is the Moon, which accompanies the Earth in its annual revolution; of which I shall first treat in the theory, and afterwards explain all her phœnomena by the lunarium, *fig. 2, plate* 11.

According to Sir Isaac Newton, this constant companion of the Earth is not of so dense a nature as the Earth is, her proportion of density to it being only as five to nine; and her mass as one to twenty-six. Her apparent diameter at her mean distance from the Earth, is $32' 12''$.

The Moon's axis is inclined to the plane of her orbit, in about $6\frac{1}{2}$ degrees, and is variable, like that of the Earth, to the plane of the ecliptic, and is effected by the same means. The plane of the Moon's orbit is also inclined to the ecliptic; the quantity of the angle between them is also variable, on account of a trifling nutation of the axis of the Moon; but at a mean rate, this angle is about 5°.

The physical cause of the motion of the Moon round the Earth and the Sun, you are already acquainted with; but her varieties remain to be treated of. These also depend on the universal effects of gravity and attraction; for, although some of them are peculiar to the Moon, and other secondary planets,—yet they are all the result of the same laws.

You must be sensible, that if the Sun acted in the same degree on the Earth and Moon, and at all times uniformly so,—their motions would be regularly performed round the centre of gravity between them, as it could not alter their respective individual action on each other; but the contrary of this supposition being the fact, and the Sun being nearest to one when furthest from the other, one must be more affected by the Sun's attraction when the other is less so, therefore their action upon each other, at different times, must by this means be variable; and it has been found excessively difficult to allow for these varieties, because the Sun's attraction does not act upon them in lines parallel to each, but in such as form an angle between them, therefore the different degree of obliquity with which it acts on each, must cause great difficulty in calculating the different aspects, &c. of the Moon, and render this physical calculation a more arduous task than any other annually entered into by mathematicians. The great Newton, it is true, has furnished the materials for mathematicians to work with; yet it must require great skill and mental labour, to complete this fabric of sublime calculation, so as to afford a true register of those facts it is to represent to the mind.

The figure of the Moon's orbit is nearly that of an ellipse; but owing to the varieties in her affections, and their inequalities amongst themselves, not regularly so.

The only equable motion of the Moon, is that with which she revolves on her axis, her revolution on which is performed exactly in a synodical month, or 29 days and a half.—This is discoverable by the spots on her disk, which prove that she always presents nearly the same face to us at the same times, though subjected to a little variation in this respect, both in regard to her equatorial and meridional parts, by which we sometimes see more of one side of her globe easterly, or westerly, and sometimes of her polar regions. These are called her librations.—The former of these appearances is occasioned by the Moon's motion on her axis being equal, and that in her orbit unequal; by reason of which, according to the difference of her situation in her orbit at the time of a particular phase, sometimes more of one side of her is seen, and sometimes more of her other side. The latter appearance proceeds from the change of place of the nodes, or different inclination of her axis to the Earth at one time than another, in the same time of her period, by which we sometimes see more of the polar region, and the parts nearer to the other, at the time of a particular phase. We see both the poles during the revolution of the Moon round the Earth, by reason of the axis of the Moon preserving its parallelism in the whole of its revolution, the same as we have observed of the Earth on the tellurium, by which sometimes one pole of the Moon, and sometimes the other, will be turned towards the Earth; but in northern latitudes, we always see her northern pole.

The path of the Moon being inclined to the ecliptic, the two points from which straight lines would be parallel to the central solar ray, are called the Moon's nodes, which are, of course, in two opposite points of her orbit. That from which the Moon ascends northerly from the ecliptic, is called her ascending node; and that by which it descends southerly from the ecliptic, is called her descending node. The line from one node to the other is called the line of the nodes; the situation of which, in regard to the ecliptic, is variable, and its motion contrary to the order of the signs, being performed in a westerly direction.

The line of the nodes is only imaginary, being nominated to express that part of the orbit of the Moon, which would, if produced, be parallel to the central solar ray.

On account of the irregularities in the affections of the Sun and Earth upon the Moon, this plane, or line of the nodes, will be affected something in the same way that the line of the nodes, or intersections of the ecliptic and equator of our Earth are, by reason of the Sun's superior power of attraction on the Moon in those two points of its orbit, arising from their being in a direct line with him.

This effect producing an acceleration of the Moon to that, arallel situation, and by degrees bringing the line of the nodes quite round in a contrary direction, which is proved by the change of situation to which the Moon's nodes are known to be subject, and which causes them to be continually changing their places in the ecliptic, in a contrary direction to their progressive motion, and to bring them round to the same point in the ecliptic from which this motion is first observed, in about 19 years.

The inclination of the plane of the Moon's orbit to the ecliptic is variable, and is affected in the same way as the nutation of the Earth's axis is; for the Sun acts on the poles of the Moon's axis also. When it is in tnat part of its orbit nearest to the Sun, the Sun must attract the Moon in a greater degree than when it is most remote from it; and if it happens to be inclined to the Sun, it will be drawn down from its natural declination by the power of the Sun; but when the line of the nodes is in conjunction with the Sun, the inclination of the axis of the Moon cannot be affected by it, because the axis is then perpendicular to the Sun's power. *Fig.* 1, *plate* 13, is a map of the Moon, which I have taken the liberty of copying from Dr. Hutton, as it is particularly illustrative, exhibiting all the remarkable spots, such as the cavities and highly illuminated parts of that orb, as seen through a telescope ; which appearances are supposed to arise from the high lands and valleys, &c. on her surface. I have also subjoined the names given to those appearances by the discoverers of them, which are those of remarkable astronomers and philosophers, or of countries, islands, or seas.

In *fig.* 1, *plate* 13, the Moon is represented at her mean libration ; the figures are for references to the names, and that spot marked with an asterisk is supposed to be one of the volcanoes discovered by Dr. Herschel.

Do not imagine that the ideas formed of mountains and valleys in the Moon are mere conjecture, unsupported by such evidences as are deducible from all the phœnomena of inaccessible objects, because I assure you they have all the visible characteristics that can be expected of objects so remote, as I will explain to you, and which your knowledge of the effects of light and shade will render intelligible.

In all situations of the Moon, the elevated parts cast a triangular shadow from the side in opposition to the Sun ; also the cavities are always dark on the side next the Sun, and illuminated on the opposite side ; exactly in the manner as we observe of light and shade on opake bodies, and those which are hollow, when made to cast shadow and appear light by any partial illumination. These appearances are found likewise to increase and diminish, according to what is perceived of any globular body turning on its axis.

In regard to volcanoes in the Moon, although it is very possible there may be such eruptions, yet the observations which are intended to authenticate this idea, are not so satisfactory, I should suppose, as those who have advanced them could wish ; because the theory is by no means sanctioned by those undeniable demonstrations deducible from the other phœnomena of the Moon ; nor will it ever be possible, in my opinion, that they should.

Various are the opinions respecting those circumstances of the Moon, that are for and against its having an atmosphere. If I may venture to give mine on a subject so ably defended on both sides, I should maintain that it has ; as I think the arguments on that side of the question are better supported than on the contrary side ; —for as Sir Isaac Newton has proved that the atmosphere of the Moon, if it has one, cannot be by one-third so dense as ours ; the density of the Moon being but one-third that of the Earth, and therefore the atmosphere on its surface weighing only one-third, its compression must be in proportion to that with which it is attracted by the Moon ; and so its effects cannot be more evident to us than they are perceived to be.

The Moon, in circulating round the Earth, according to her situation in respect to that and to the Sun, presents more or less of her illuminated face to us ; but at one part of her orbit it is wholly turned from us.

From the different phases she exhibits to us we also know that her form is globular, although from the laws of centrifugal and centripetal forces, we know that it cannot be perfectly spherical.

When the Moon is in opposition to the Sun, or in the opposite sign to him, she turns the whole of her illuminated side towards the Earth, and is seen by us on our meridian at midnight. She is then said to be full, appearing as represented at A, *fig. 2. plate* 13, in the ecliptic circle E F G H, to which the place of the Moon is transferred by a spectator on the Earth ; because it moves nearly in that apparent path. S is the Sun in his proper situation, in respect to the Earth, at the time of full Moon ; and A B C D the Moon in four different and opposite points of her orbit.

As the Earth at T moves easterly in her orbit, the Moon changes her figure as viewed from the Earth, the western part of her illuminated surface declining from the Earth, in her departing from her opposition to the Sun; and when she is in quadrature with the Sun, or only as at B, three signs from the place of the Sun, as viewed from the Earth, she appears like a crescent, on our meridian, at six o'clock in the morning. From this place, as she advances easterly towards her conjunction with the Sun, she diminishes in appearance to us, because more and more of her enlightened face is turned from us, by her now getting between the Sun and us; and when in conjunction with him, as at C, the whole of her illuminated face is turned from us, she being in a direct line between the Earth and the Sun. This obscuration continues about four days, when she again begins to shew a part of her illuminated surface; and we see her on an evening, soon after sun-set, in the shape of a fine crescent. When she is in conjunction with the Sun, we say it is New Moon. From her conjunction at C, as she is advancing towards her quadrature at D, she is continually increasing in her apparent size; and when she has arrived at D, we again see her on our meridian at six o'clock in the morning, and exactly half full, or bisected; from the time of this quadrature, till she again arrives at her place of opposition to the Sun, or at full, as at A, she is continually increasing in size, till Full Moon again at T.

These changes she goes through in the course of 29 days, 12 hours, and 44 minutes, which time is called her lunar synodical month, in contradistinction to her periodical lunar month,—the former being the time of all her changes of appearance as referred to the Earth,—and the latter (which is performed in 27 days, 7 hours) being the time of performing her revolution round the Earth. The period of her being in conjunction with the Sun, as at C, and her return to it again, is also called a lunation.

The places of the Moon, when she is at new and full, or in con-

junction and opposition, are called the syzygies. From either of
the syzygies to the quadrature, the gravity of the Moon towards
the Earth is continually increasing; and the attraction of the Sun
also acting contrary to her progressive motion, these two forces
counteract each other in a degree, by which the motion of the
Moon in her orbit is continually retarded; but from the quadratures to the syzygies, the Sun and Earth conspiring in their effects,
the motion in her orbit is perpetually accelerated.

These varieties of the affections to which the Moon is subject,
occasion her orbit to deviate something from the form of a regular
ellipse, although it is rather oval.—The Moon's diameter is to that
of the Earth, as 20 to 73, being computed at 2,180 miles. Hence
the surface of the Moon is to that of the Earth, as 1 to 13; so that
the Earth must reflect 13 times as much light upon the Moon, as
the Moon does upon the Earth.

The Moon cannot have any great variety of seasons, as her axis
is almost perpendicular to the plane of her orbit.

The sloping ring, Q D, of the instrument, *fig.* 3, *plate* 10, is to
represent the plane of the Moon's orbit, which should be inclined
at a mean rate in an angle of 5° 18′ to the ecliptic plate of the
orrery, as shewn by the plane of the nodes ☊ ☋, which is distinguished by a brass wire fixed from A to B. In this instrument, the
line of the nodes has the motion I have previously mentioned, contrary to the order of the signs, as may be seen by comparing its
situation at different times, in respect to the signs and degrees described on the plate under it, which represents the ecliptic. These
also shew that the Moon cannot be exactly in the plane of the
ecliptic, except when she is in her nodes; as in all other parts of
her orbit, she goes either north or south of that path, which deviation from it is called her latitude, and does not exceed about
5° 18′ on each side of the plane of the ecliptic, or central solar
ray, as mentioned above, which are called her highest and lowest

situations in her orbit, and which are seen on this ring, these degrees and minutes being engraved on it.

That part of the plate RS, under the Earth and Moon, which has the names of the signs engraved on it, is moveable in a socket; so that in working problems by this machine, those signs may be made to correspond with the large ecliptic of the orrery; thus— Whatever sign the Sun should appear in from our Earth, on the large plate, that sign of the small plate should be placed next the Sun, because both the Sun and Moon are transferred to that circle in the Heavens. On the outside of the signs, on the small ecliptic circle of the lunarium, are engraved the days of the Moon's synodical, and also her lunar month; by comparing which with the signs and degrees on the small ecliptic plate, (those two circles moving independently of each other) the difference between the periodical lunar month, and the synodical lunar month, is clearly seen, as well as the cause of that difference. The ellipse X Y is to shew the Moon's elliptical orbit, on which are marked apogee and perigee, to denote the Moon's greatest and least distance from the Earth.

I shall explain the difference between the periodical and synodical month, first by a diagram, and afterwards by this instrument.

The difference between them depends on the progressive motion of the Earth, together with that of the Moon, round the Sun, compared with their motion round the centre of gravity between themselves. The centre of gravity between them is exhibited by the pin at Q, being as much nearer to the Earth, as the Earth is larger than the Moon. If these two bodies had not a progressive motion round the Sun, the synodical and periodical months would be of the same length.

Let A B C D, *fig.* 3, *plate* 13, represent the great ecliptic, to which the place of the Sun is transferred, and 1234 the small ecliptic, to which the place of the Moon is transferred, and T O the Earth in its orbit. First, suppose the Earth to be in her orbit at T, and the place of the Moon to be at ι, in the line T ι A; then, as the Earth and Moon are performing their periodical monthly revolution round the centre of gravity between them, which is performed in twenty-seven days and a half, they will also advance in their annual revolutions round the Sun; and therefore, when they have performed their periodical months, they will have advanced a whole sign in the ecliptic, to O. But the Moon will not again appear in conjunction with the Sun, or in a right line with the Earth and Sun, until she has described the arc E F, which is wanted to complete her synodical month, or time of being in the same sign with the Sun in respect to the Earth, the difference between that and her periodical month being the quantity of that arc. I shall now illustrate this by the lunariun.

I will place the Moon in conjunction with the Sun, as in the diagram, and fix the ecliptic plate under the Earth and Moon, by the large one, so that the place of the Sun and Moon, as supposed to be seen from the Earth, may be in the same sign and degree of the ecliptic, letting the days of the month under the Earth and Moon correspond with those set down in an Ephemeris, or as exhibited by the large plate of the orrery.

On turning this instrument, you will perceive that the Moon moves in an orbit inclined to the ecliptic, and that it turns once round upon its axis whilst it is performing its synodical month, and that it will return to the same point of the small ecliptic, or perform its periodical month in twenty-seven days and a half; but will not again appear in conjunction with the Sun in less than twenty-nine days and a half, as may be seen by the days of the month under the Earth and Moon; the annual index will point out the cause of this, which is, the Earth and Moon having advanced nearly a whole sign in their annual orbit round the Sun in that time. The Moon's motion round the Earth is from west to east, and the apparent annual motion of the Sun is also from west to east, being both in the same direction, but with different velocities, the Earth getting the start of the Moon, in respect to her situation with the Sun, which requires her to travel two days longer in order to overtake, or again be in conjunction with the Sun and Earth.

This instrument shews the difference of the face of the Moon, exhibited at different times, for the same phase, on account of the revolution of the nodes, but not that difference occasioned by the obliquity of its axis, as it is always perpendicular to the plane of its orbit, on this instrument, contrary to fact. The different phases of the Moon are also pleasingly illustrated by the lunarium; and the places of the nodes exactly correspond with those perceived throughout any space of time, if once made to answer to their proper stations and times; as also the Moon's age and phases for any length of time, by one adjustment, to those which take place at any given period.

Having examined into some of the phœnomena of the Moon, I shall postpone what remains to be observed of it till my next Lecture, and conclude this by referring to the foregoing part of it, in which we treated of the variety of the seasons. Every reflecting mind that minutely examines into the causes of the vicissitude of the seasons will not stop there, but will proceed to examine into the benefits arising from them, and the inferences deducible from those benefits. For being sensible that the whole course of nature is intended to convey instruction to rational beings, it next compares, then judges, and finally is amply repaid, by the conso-lations deducible from the minutest circumstance in nature.

Thus, perceiving the Sun to be the chief natural agent em-ployed in regulating the seasons, we next discover that his power is circumscribed by such laws as render his agency most efficacious to us; and although acting as an agent, yet that he is subject to a controling infallible power; from which observations we are led

to judge of the benevolence and impartiality of all the dispensations of the Deity; deducible from the above, and from observing
that He bestows on each portion of his creation an equal portion
of his blessing.

Independant of the constitutional advantages afforded animal and
vegetable nature, by the vicisstude of seasons, we may rejoice in
them on account of the vigour of our mental enjoyments; which
are thus insured and continued to us. If we had perpetual sun-
shine, it would cease to afford us the delight we now experience in
its return to us, after a temporary absence of its enlivening influence.

With what charms is the vegetable creation decked in spring,
which delight us as much by their renovation as their real beauties! succeeding the total deprivation of such grateful objects, they
communicate a pleasing sympathetic renovation in all sensible
minds—exhilarating the spirits, thus animated by their gay and
lively influence. What is there in art or invention that compen-
sates for the charms of nature?—Those native beauties, ever new
and varying, always serve to delight the eye, improve the fancy,
and cheer the heart.—Happy is it for those who have so improved
a taste for rational studies, so to fill their time at the period of
temporary decay, and deprivation of objects so charming, as to
escape from the vain delusions which destroy the health and ener-
vate the minds of thousands, by which they are prevented from
relishing the benefits and sweets of returning spring. Whereas
those who have cultivated their minds, and stored them with useful
knowledge, will, at the return of each vernal period, ever taste new
delights—and when autumn deprives them of the charms of nature, they will enjoy them in prospect, and by considering them
rationally, look forward to their return with pleasing satisfaction.

Who can behold the renewal of all nature in spring without inferring their own resurrection, of which it is so striking an em-

blem? For as the tree sheds its leaf in autumn, so shall our mortal part die away; and as that renews its foliage in spring, so shall we flourish again. Thus we perceive it to be with all the vegetable world, and also with many insects. One instance this moment presents itself, which I think a striking figure of our present degenerated, and also of our future exalted, state, although evinced in the humble caterpillar.

This insect changes its natural state of existence in a manner analogous to our tranflation from this life to a better—for after being to all appearance dead, it rises again in a new and beautiful form. In the winter of its age, foreseeing, we may suppose, its approaching change, it begins to prepare its tomb, and works unremittingly till it is shut out from all visible means of subsistence; in this state it continues a certain time, and then rises again from its temporary obscurity to a life of joy; in which new and beautiful form it wings its way with exalted renovation, soaring above those things which in its former degraded state appeared its proper sphere of action—relishing only the most refined pleasures.

How emblematical is this of our existence! We now find pleasure in the gratification of those senses our bodies are endowed with, which hereafter we shall lose, with them, all relish for. Like this insect, we must pass our allotted time in our present state, then be entombed, and to all appearance dead—yet shall that Power which supports the chrysalis in its state of apparent inanity, sustain our spiritual part,—and finally, at the time decreed by divine command, we shall rise again to a most pure, a most ennobled state, in which our present body shall not appear, but we shall be so cloathed as to fit us for the participation of those enjoyments which the refined nature of our situation shall require. Therefore the decay of our mortal part, instead of being a subject for regret, if we improve our mental enjoyments in proportion, conveys the greatest consolation to a reflecting mind—by the assurance it gives us of a better state than this; first, by the necessity of a new form for the enjoyments it promises; and, secondly, by a consciousness of having a mind capable of such participations. And the inference we may justly and finally make upon the consolation this change of form brings with it, is, that were we to rise from the grave incumbered with our bodily infirmities, we should be unfit for a purer state, or more exalted gratifications.

Such are the consoling testimonies perpetually arising in the contemplative mind; that investigates nature for the most useful purposes—to strengthen the judgment—fortify the mind—refine and exalt the ideas—rectify, expand, and ennoble the heart.

Thou, Power Supreme, by whose command I live,
The grateful tribute of my praise receive.
To thy indulgence I my being owe,
And all those joys which from that being flow;
Thy skill my elemental clay refin'd,
The vagrant particles in order join'd;
With perfect symmetry compos'd the whole,
And stamp'd thy sacred image on my soul;
A soul susceptible of endless joy,
Whose frame nor force nor time can e'er destroy;
Which shall survive when nature claims my breath,
And bid defiance to the darts of death;
To realms of bliss with active freedom soar,
And live when earth and skies shall be no more.
Author of life! in vain my tongue essays,
For this immortal gift, to speak thy praise!
How shall my heart its grateful sense reveal,
When all the energies of words must fail?
O may its influence in my life appear,
And every action prove my thanks sincere!   Mrs. E. CARTER.

END OF THE NINTH LECTURE.

Plate XIV.

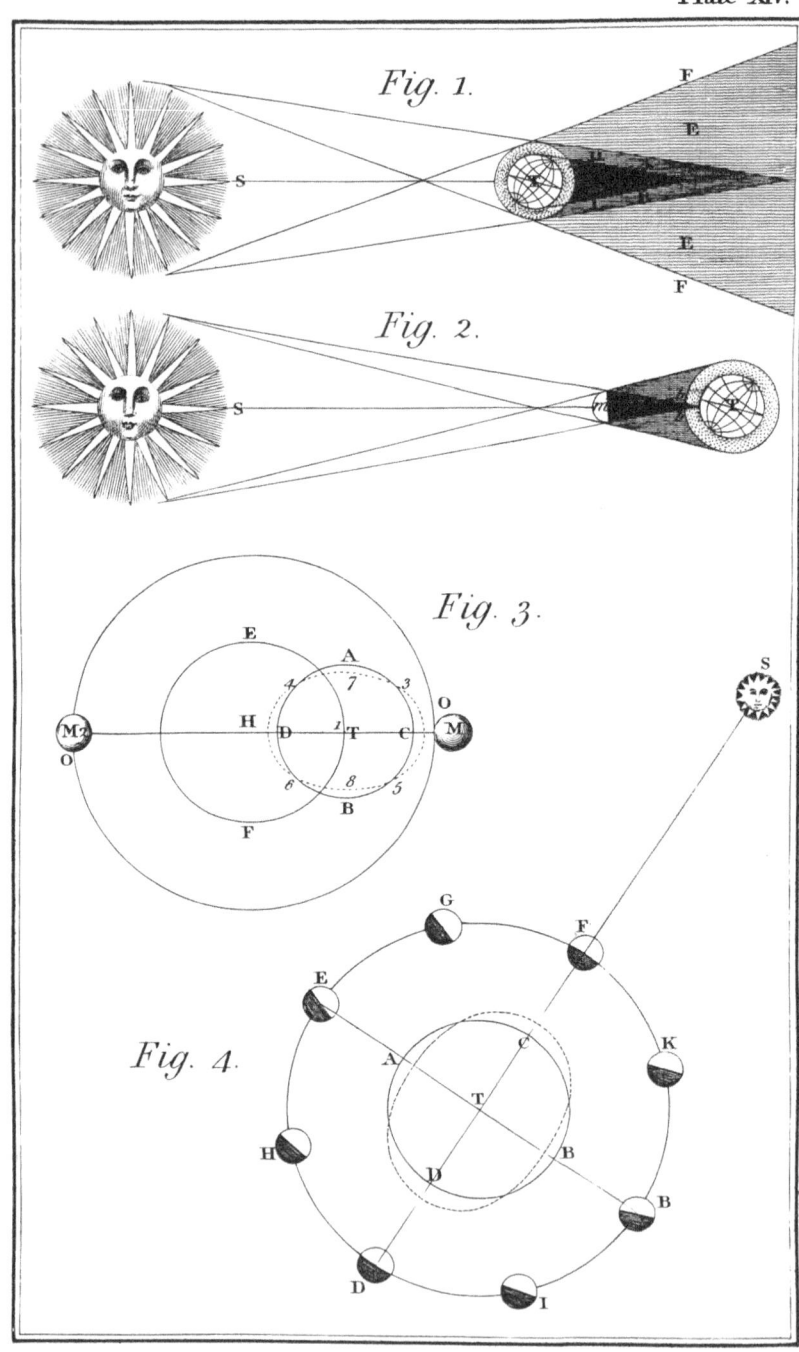

Fig. 1.

Fig. 2.

Fig. 3.

Fig. 4.

M. Bryan delin.ᵗ

T. Conder sculp.ᵗ

LECTURE X.

OF THE LAWS OF SHADOW.—OF ECLIPSES.—THE PHŒNO-
MENA OF THE TIDES CONSIDERED; AND HOW PRODUCED.
—CONCLUDING ADDRESS TO MY PUPILS.

———— Was ev'ry fault'ring tongue of man,
ALMIGHTY FATHER! silent in thy praise,
Thy Works themselves would raise a gen'ral voice,
Ev'n in the depth of solitary woods
By human foot untrod; proclaim thy pow'r,
And to the choir celestial Thee resound,
Th' eternal Cause, Support, and End of All!   THOMSON.

PREVIOUS to my acquainting you with the phœnomena of
eclipses, it will be necessary to say something of the laws of shadow.

Every illuminated object must cast a shadow in a direction parallel to the rays of light by which it is illuminated. The more
intense the light the stronger will the shadow appear, because the
contrast is the greater; not that the shadow is of itself darker, for
all perfectly opake bodies will cause the same degree of shade.

If the luminous body and the body illumined are of the same
size, and parallel to each other, the shadow cast by the latter will be
exactly of the same dimensions with the former. But when these
two bodies are unequal to each other, and the luminous body, for
instance, is larger than the opake one, then the shadow always becomes less and less, as it recedes from the body; and when the
opake body is larger than the luminous one, the shadow increases
the further it goes from the body—the rays diverging or spreading
continually in the latter case, but converging or approaching
continually in the former.

F f

That faint shade which is seen between the dark shadow, cast by an object, and the light, when the luminous body is larger than the opake one, is called a penumbra. It arises from the larger and luminous body sending some of its rays into those parts, in which its direct influence cannot dart, on account of the interposed body. The penumbra will be darker or lighter gradually, from the extremity of the diameter towards the dark shadow subtended by the opake body, so that the outer part will be lightest, because it will receive more of the Sun's rays than that part which is nearly behind the opake one; and as the shadow of the opake body becomes gradually less and less, the penumbra grows gradually broader and broader. These positions you may understand by looking at *fig.* 1, *plate* 14, in which S is intended to represent the Sun, and M the Moon. You perceive that the latter is much less than the former. Suppose C to be the dark shadow of the Moon, received on a plane at that distance, as at M, it would be diminished in its apparent size to the diameter AB; but as some of the rays from the Sun must enter between the lines ET and EF, they will cause the shadow to be fainter in those parts.

The word eclipse is adopted to express these deprivations of light, and is derived from a Greek word which signifies a fainting or dying away, and is therefore very properly expressive of what happens to the Moon by the intervention of the Earth between it and the Sun; although not equally so when used to signify the circumstance of the intervention of the Moon between us and the Sun. For, in the former case, the Moon is absolutely deprived of the Sun's light; but in the latter, the Sun loses nothing of his wonted splendor. Yet, as the same term is always used to express both circumstances, we must conform to the popular mode.

When the Moon passes through the shadow of the Earth, being deprived of the Sun's light, she becomes invisble to us, or rather in such parts of her only as pass through the Earth's shadow; as, on account of the obliquity of the Moon's orbit to the orbit of the Earth, she is not totally eclipsed; that is, her whole diameter does not pass through the shadow of the Earth, excepting when these obscurations happen at the time of the Moon being in or near the line of the nodes. When the Moon is exactly in the line of the nodes, she is said to be centrally and totally eclipsed, because her centre passes exactly through the axis of the Earth's shadow. When the Moon is within 12 degrees of the line of the nodes, at the time of her being full, she will be partially eclipsed; but when she is more than 12 degrees from it, at the time of full Moon, she will pass clear of the Earth's shadow. Thus you perceive the reason why the Moon is not eclipsed every time of her being at full; and that this is occasioned by the plane of the Moon's orbit being inclined to that of the Earth; whereas, if these two orbits were in the same plane, the whole of the Moon's diameter would pass through the shadow of the Earth every time she was in opposition to the Sun, and she would be totally eclipsed at every full Moon; by which the great advantages procured by her superior illumination at those periods, would be lost to us.

If the Earth and Sun were both of a size, the shadow cast by the Earth would continue the same size with itself, to any distance; and therefore the remotest planets of the system would be involved in obscurity as well as the Moon; which not happening to the superior planets, Mars, Jupiter, Saturn, or the Georgium Sidus, we are certain that the Sun and Earth are not of the same size: also, that the Earth must be less than the Sun; otherwise its shadow, by continually extending itself, would, as in the former case, reach the orbit of the most distant planet in the system. Thus we are certain, by these as well as every other possible deduction, that the Sun is larger than the Earth.

The Earth being less than the Sun, and of a spherical form, the

F f 2

shadow it casts must be a cone, terminating in a point at a certain distance from it, and falling upon that point directly opposite to the Sun's place in the ecliptic.

A lunar eclipse, which is the one I have been particularly describing, is in the same degree to all the inhabitants of the Earth; only observed at different hours, at different places, according to the longitude or meridian of each place. The shadow appears fainter on those parts of the Moon by which she makes her ingress and regress; because the shadow cast by the atmosphere of the Earth is fainter than that cast by the Earth itself, as H I, *fig.* 1, *plate* 14.

A total and central eclipse of the Moon is represented in *fig.* 1, *plate* 14; the Sun, Earth, and Moon being all in a direct line, as the Moon is at the time of the nodes.

Solar eclipses, as they are called, can never happen but when the Moon is in or near the line of her nodes at the time of new Moon, or of her being in the same sign with the Sun. These eclipses, like those of the Moon, are either total or partial, and are occasioned by the shadow of the Moon cast upon the Earth.

The Moon being less than the Sun, subtends a conical shadow; and as she is less than the Earth, only a small part of it can be involved in her shadow; so that eclipses of the Sun, whether partial or total, are not at all parts at the same time, but to a small portion of the Earth only, and which is supposed to be no more than 180 miles broad at the greatest, or a total eclipse of the Sun.

When the Moon is more than 17° from her nodes, at the time of new Moon, she hides only a very small portion of the Sun; but the nearer she is to them, the greater will be his obscuration; and when in the nodes, the Sun will be totally eclipsed. The beginning and end of an eclipse of the Sun, will happen at different times at different places; and the degrees of it will appear differently to different parts of our globe. To some it may appear partial, to others total, at the same time. That part of the Earth which is in a direct line with the Sun and Moon, will have a greater eclipse of the Sun, than those parts which are furthest from that central line between them.

In total eclipses of the Sun, the limb of the Moon is surrounded by a pale circle of light, which, on first entering on its passage over that luminary, is perceived to have a vibration, as also in its departing from it; a circumstance produced to authenticate the idea of the Moon having an atmosphere; which appears to me an indisputable fact, although much has been advanced to contradict that idea; such as its always appearing of the same uniform brightness when our atmosphere is clear: whereas, if she has an atmosphere, they suppose this would not be the case; which conclusion proceeds, I imagine, from not considering the extreme rarity of the Moon's atmosphere, which would not permit those evidences of its existence required by them. Others maintain, that as no seas nor lakes are in the Moon, there can be no atmosphere, being no water to be raised in vapour. Even supposing there is no water on the surface of the Moon, as some have inferred from the external appearance of that body; yet, I should suppose that there may be a subterraneous store for the purposes of vegetation; and that from thence it might be raised by the heat of the Sun, in the same manner as the subterranean waters of our Earth.

Independant of my observation on it, others have denied the positions from which the idea of the Moon not having an atmosphere has been deduced. They have asserted, that the Moon does not appear always of the same brightness, even when our atmosphere is clear; but that her brilliancy is also affected by the different state of her own atmosphere, at different times. Lastly, the Moon's atmosphere being so much less dense than ours, it may not cause any sensible refraction of a star, viewed through it from the Earth; which is another observation made by those who will not allow the Moon to have an atmosphere. On summing up the arguments, for and against these circumstances of the Moon, I think we may naturally conclude — that the Moon has an atmosphere, and is not destitute of water.

Fig. 2, *plate* 14, represents the situation of the Sun, Moon, and Earth, at the time of a total and central eclipse of the Sun : S the Sun, M the Moon, and T the Earth ; also, *b b* the penumbra, and *a a* the dark shadow of the Moon. The Moon's atmosphere is too rare to cast any shadow, and therefore is not noticed in this diagram.

If the Moon's nodes had not that retrograde motion which I have explained to you, they would be in conjunction with the Sun every six months ; and therefore there would be a regular return of the same eclipse in that time. But the eclipses happen so much sooner each year than the preceding, as proves this motion of the nodes, and the quantity of it ; so that it is known that they have a retrograde motion, and that the period of their retrogression through all the signs of the ecliptic is performed in about nineteen years.

Again, if in nineteen years there were exactly a certain number of courses of the Moon, this would be the period of their returning, according to their aspects ; so that we should have exactly the same eclipses as had happened in the former nineteen years. But this is not the case, as the Moon performs thirty courses and a quarter whilst the periodical retrogression of her nodes is completing. However, this difficulty has been surmounted by remarking the time in which 223 courses of the Moon are performed ; which is found to be eighteen years, eleven days, seven hours, and forty-three minutes ; after which period, the conjunction of the Sun with the same nodes occurs. Thus a period is obtained, in which we know that the same eclipses will happen again, although not exactly at the same time.

The most common number of eclipses in the year, are two of the Sun and two of the Moon; yet in some years there are six, and in others only two, on account of the disagreement between the periods of the eclipses and those of time.

As the apparent diameters of the Sun and Moon are nearly equal, the duration of a total eclipse of the Sun never exceeds two minutes; but the same eclipse of the Moon may continue above three hours, because the shadow of the Earth, where the Moon passes through it, is near three times as large as the diameter of the Moon.

As eclipses happen always in those months in which the Sun enters the signs where the Moon's nodes are situated, and the nodes being directly opposite to each other, the time of eclipses will occur at about six months distance from each other; so that if one happens in January, there will be another in July following.

Tables are constructed for the purpose of calculating eclipses for any length of time, by allowing the three minutes eleven seconds that the nodes of the Moon revolve in a retrograde direction each day; and as in 18 years, 11 days, 7 hours, 43 minutes, the Earth and Moon advance each just as far beyond a complete revolution in the ecliptic, as the nodes want of completing their retrograde one, they will meet at that time, and be in the same respective situation to each other that they were in 18 years, 11 days, 7 hours, 43 minutes before.

Astronomy and Geography have both been much benefitted by the observation of eclipses. In the former, they have authenticated the idea of the comparative sizes of the Sun, Moon, and Earth, and the period of the Moon's retrogression; and in the latter, the rotundity of the Earth, and the longitude of places on it, have been also ascertained by them.

What I have advanced respecting eclipses of the Sun and Moon, may be very pleasingly exhibited by the Lunarium. On turning the Earth and Moon on in the Earth's annual orbit, you will perceive that the Moon and the Earth are not in a direct ine with the centre of the Sun, except at the time of the Moon's nodes, and when in opposition and conjunction; therefore, when the Moon is at new or full, at the time of the nodes, there will be in the former case an eclipse of the Sun, and in the latter an eclipse of the Moon. The period of these eclipses being known for any time, if we fix the line of the nodes in their proper situation, for either of them, and turn the Earth and Moon on in the Earth's annual orbit, we shall see when all the rest will happen, through any space of time. If we continue our observasions for the whole 18 years 11 days, we shall perceive that there will be 223 courses of the Moon in that time; at the expiration of which period, the Sun and Moon will be in conjunction, at the same node and in the same sign they were in 18 years 11 days before; and shall also be able to ascertain every eclipse of the Sun and Moon which may occur in the intermediate time.

I shall now dismiss this subject, having, I think, explained it very fully; and shall proceed to consider those affections of the Sun and Moon, which cause the flux and reflux of the waters of the ocean.

The rising of the tides is occasioned, in one respect, by the same principle as causes the Earth to be an oblate spheriod, but not in all; the concurring circumstances are these — the velocity of their external parts being increased by the extension of those parts, and their being acted upon by two forces.

But to be more particular. The surface of the Earth being more than half covered with water, and a fluid being more easily drawn from its situation, by any power, than a solid — the fluid parts of the Earth must be more affected by the attraction of the Sun and Moon, than the solid parts of it. This causes the tides in our seas;

and it is also well known that they are principally produced by the attraction of the Moon, as they are found to correspond with her periods, and to be in proportion to them.

The Sun's power in raising the tides must be but small in comparison to that of the Moon, on account of his great distance from the Earth. In fact, it may rather be considered as counteracting the effects of the Moon, in certain circumstances of it, than as occasioning any sensible flux or reflux.

Kepler was the first who produced a rational theory of the tides, and Newton has confirmed it on the known principles of gravity or attraction.

The sea is observed to ebb and flow alternately twice in twenty-four hours forty-eight minutes; rising continually for about six hours, and falling in the same space of time.

The time of high water at any place, is about forty-eight minutes later every day; because the Moon is twenty-four hours forty-eight minutes from being on and returning to the meridian of a place in one day, which is called a lunar day. The sea flows, or it is high water, always at the time of the Moon being nearly on the meridian of any place, or in the point of the ecliptic opposite the meridian, below the horizon. The sea ebbs, or it is low water, whenever the Moon passes nearly the horizon of any place, either in rising or setting to it.

The waters of any place are raised rather higher when the Moon is on the meridian of a place, than when she is on the opposite meridian; the cause of which you will perfectly underſtand when I treat of the physical causes of the tides.

The Sun always raises the waters a little when he is on our meridian, and when at the opposite one; but his tides are not distinguished from the others, because they are made evident only by diminishing them;—when acting in opposition to the Moon's attraction, they produce what are called neap tides; and when in

G g

conjunction with it, spring or high tides: so that at the time of new and full Moon, we have high tides; and at the quarters, low tides.

The attraction of the Moon must, as well as that of the Sun, affect the equatorial parts of the Earth more than those which are nearer to the poles, both on account of its less obliquity to the equator than to the poles, and also on account of the increase of matter about the equator.

The degrees of the comparative effects of the Sun and Moon on the tides, are not the same every day, according to the periods of the Moon, or the relative situations of the Sun and Moon to each other; because the lunar and natural days do not correspond in time, the lunar being about twenty-four hours forty-eight minutes, and the natural or solar day about twenty-four hours.

The tides are equal to each other at each returning situation of the Moon, in respect to the Earth and Sun: therefore, when the Moon is in conjunction with the Sun, we have the highest tides of all; when she is in opposition to him, nearly as high as when in conjunction; but when in quadrature with him, we have the lowest tides of all, because they act cross-ways on the Earth. When the Sun and Moon are in the syzygies, or in opposition and conjunction, at the time of the equinoxes, we have the highest tides of all, because then the Sun is in the plane of the equator, and the Moon the same, or very nearly so. But as the Earth is nearer to the Sun in winter than in summer, these tides happen rather before the vernal equinox, and after the autumnal. The degree of these tides is variable, as they depend on the situation of the Moon's nodes at those times, and the time of its being new or full Moon.

The mean comparative forces of the Sun and Moon attracting the waters, are supposed to be as 1 to $4\frac{1}{2}$; therefore, the total effect of the Moon's attraction must be in proportion to the difference. The action of the Sun has been supposed sufficient of itself to raise the waters of our seas about two feet, whence that of the Moon may, at a mean rate, raise them nine feet; that is, at a medium. From this we infer, that at spring tides, the water will be raised eleven feet by the Sun and Moon acting conjointly; and at neap tides, or when they are acting cross-ways to each other, seven feet; allowing for what is added to the individual power of the Moon in one case, and what is subtracted from it in the other, by the action of the Sun.

I shall now apply these observations, and illustrate them by diagrams. In *fig.* 3, *plate* 14, let M represent the Moon in her orbit OO; ABCD the Earth in it's orbit EF. Let the line from D to M be considered as that of their attraction, by which they are made to revolve round the centre of gravity at H; and let this centre of gravity be supposed 40 times nearer to the Earth than to the Moon, as it is in fact, on account of the Earth containing 40 times as much matter as the Moon.

Every body moving round a centre, endeavours to fly off from that centre by its centrifugal force; and as that part which is furthest from that centre moves faster than that which is nearest to it, this causes that part of the Earth which is opposite to the Moon, to exert a greater force from its centre than that which is nearest to the centre of gravity, or facing that and the Moon; and of course, causes the tides on that part to rise by the centrifugal force, and those facing the Moon by the Moon's attraction or centripetal force. Without allowing for the centrifugal force, we could not account for the tide rising on opposite meridians at the same time; for the power of the Sun, even when on the opposite meridian to the Moon, would of itself not occasion any considerable degree of tide: but these two forces, having been allowed, account for the water being raised on both sides of the Earth at the same time, as at C and D.

As the nearer any part of the Earth is to the attracting power, the more it must be affected by it, we readily conceive that the part of the Earth at D must be more affected by the Moon's attraction than the parts C A B, or the centre of the Earth T, and will of course cause it to approach the Moon, in the manner and in the proportion as from 4 to 6.

By whatever means, and in whatever degree, the waters of the sea are raised in one part, they must be proportionably depressed in another; so that when the Moon raises the waters at 4 and 6, and the centrifugal force at 3 and 5, they must be reduced at 7 and 8.

The waters at 3, 5, and 4, 6, are raised to the same height, because action and re-action are equal, in proportion to the quantities of matter; and therefore the centrifugal force on one side of the Earth (occasioned by its moving round the centre of gravity between itself and the Moon) occasions the waters on the Earth opposite to the Moon, to be raised in the same degree, as the Moon's attraction causes them to be raised on the other side. For as these two forces balance each other, the fluid parts of the Earth must be equally affected on both sides of it. You know, that if it were not for this centrifugal force, the Moon would fall into the Earth by the superior attraction of the latter, as I have before observed of the planets and the Sun, it being the same in all of them. The centrifugal and centripetal forces counteracting each other, all bodies revolving round a centre will keep at a proper distance from each other, preserving an equilibrium; by which means such effects as would be destructive to harmony and earthly existence, are prevented.—Thus, we find, all things regulated by weight and measure!

The effects of the centripetal and centrifugal forces on the seas of our globe, may be familiarly illustrated by any flexible body suspended from a cord, and whirled round by the hand. The hand, or the centripetal force, being considered the same as the Moon's attraction, and the force the body exerts from the centre, or its centrifugal force, will cause the body to swell out in two opposite parts—the one next the hand, and the other opposite to it. These effects exactly correspond with those produced by the Moon's attraction, and the centrifugal force, on the waters of the Earth. You will also observe that the flexible body, in proportion as it extends itself in those two directions, will become depressed on the two other sides of it, the same as the waters are on the parts of the Earth 7 and 8.

What are called neap tides, (which are those in which the water is not raised quite so high as in the illustration I have given you of high and low tides) are occasioned by the Sun and Moon acting cross-ways to each other; as thus:—Suppose ABCD, *fig.* 4, *plate* 14, to be the Earth, and T its centre; from which you perceive four lines are drawn, one to the Moon at E, another to the Moon at F, and through the latter to the Sun at S; a third to D, and another to B.

When the Moon is at the quadrature, or 90 degrees from the Sun's place in the ecliptic, as at E, her attraction on the waters at A and B will not be sufficient to raise them so high as when she is at F; because, when the Sun and Moon both attract in the same direction, as at F, they will raise the water higher than when they act cross-ways to each other, as they do when the Moon is at E, and the Sun at S.—Therefore the latter is called neap or low tide; for although the waters will be raised in those parts and at those situations of the bodies, yet at a mean rate they will not be so high by one-fourth. But when the Sun and Moon act conjointly on the waters, as at C, F, S, then the waters will be raised one-fourth more than a mean or common tide; and these are called spring tides—the depression in one case, and the elevation in the other, being in proportion to the effects of the two bodies, acting in opposite directions on the Earth. And as the effects of the tide are on the part of the Earth at C, such will they be on the opposite part D, for the reasons before given.

When the Moon is at G and K, or 45° from the Sun's place in the ecliptic, being at the mean distance between the place of the highest and lowest tide, the tides are at a mean rate.

The height of the tide, at all times, must be regulated by the circumstances of the place; as also the time of its being at its height: because, for example, if the inlet is narrow, it will rise higher and quicker than when it is broader, from the opposition of the shores to the waters, which is the reason that they rise but little in the open ocean.

The highest tide does not happen when the Moon is exactly on the meridian of a place on the Earth, or on the opposite meridian to it, but some time afterwards; because any fluid receiving a certain impulse, will continue its motion, in the direction of its impulse, after the power has ceased. Thus the Moon's attraction causing the waters to rise to a certain height, will, on account of the motion communicated, cause it to rise higher; so that its tides are higher than accounted for by attraction only, and are at the highest about three hours after the Moon is on the meridian, or opposite meridian, of any place;—for although the impulse has ceased, yet the effects of it are continued.

The air, or atmosphere of the Earth, being lighter than water, must be affected by the Moon's attraction, and in a greater degree than the waters are. The effect of the Moon's attraction on our atmosphere, is one cause of the high winds about the time of the equinoxes; but the change of its temperature is the principal cause.

As all that relates to the different temperature of our atmosphere seems connected with the theory of the Earth, I will endeavour to explain the nature of its agitation at different times, and in different parts of it*.

You are already acquainted with the expansion of air by heat,

* I shall not go into a physical consideration of the air, being foreign to my present purpose, and as I have already delivered two Lectures wholly on that subject for the information of my pupils.

and the contraction by cold, to which it is subject; also with the effects produced on the animal constitution, by the changes in its temperature and weight; I shall, therefore, proceed to inform you, how these effects are produced.

When the air, in one region of the atmosphere, is by the attraction of the Moon, or any other cause, dilated, it must, by that expansion, force itself into another, which, if it happens to be less rare, will occasion a wind or agitation. But the principal cause of the extension of the air, is heat; so that when the Sun, or rather the reflection from the Earth, has heated the air to a certain degree, it becomes lighter or more rare in those parts than in the adjacent, which causes the colder or denser air to rush into it, in order to restore the equilibrium of weight and pressure. The rushing of the colder air into the rarer, causes that agitation we call wind; and when the states of the different regions of air are in a great degree unequal to each other, and proceed from different quarters of it, on account of the heterogeneous particles with which they are loaded, on rushing into contact with each other, they cause tempests and hurricanes.

But these apparent discords of nature produce the most harmonious and beneficial effects—purifying the air from all noxious vapours—dispersing the contaminated particles, and thereby preventing an accumulation unfavorable to animal existence: and thus an influx of fresh air is poured into all regions, so as to produce an equilibrium, and to sustain a healthful temperature in all;—by moderating the heat of the parts and regions most dilated by the Sun, as the southern with an influx of air from those which are colder, and meliorating those of the colder or northern regions by the warm air of the southern. The former of these effects is most sensibly felt by the inhabitants of the tropics, on whom the Sun's beams fall with accumulated ardour, which cause the air in those parts to be much heated during his influence. Yet these effects being less sensible on the surface of the waters of the ocean, by which the tropics are surrounded, on account of their absorbing a great quantity of the Sun's rays,—when the Sun sinks below the horizon of those islands, the air regains its equilibrium : The cooler air of the ocean rushing into the heated air, refreshes and purifies it by its coolness and wholesome nitrous impregnation.

Within the torrid zone, an easterly wind blows the whole year, the constancy of which is very useful to navigators ; for which reason it is called the trade wind. The cause of this wind has been differently defined ; but the most rational mode of accounting for its uniform action, is the following : The regions between the tropics being more heated, on account of their less obliquity to the Sun, and receiving more of his rays than any other part of our globe, and the atmosphere in those parts having acquired an extension by the heat ;—on the Sun's departure from any spot, the colder air will rush in, as it were, to restore the equilibrium. This may occafion the constancy of the wind in question. Its direction may be accounted for thus : The diurnal motion of the Earth being from west to east, the apparent motion of the Sun is in a contrary direction, and thus the Sun passes it, as it were, westerly. That part of the atmosphere which is at any time subjected to his influence being most heated on his retreating westerly from that part, the colder air rushing in behind, as I before observed, will cause a constant succession of wind from the east in those parts.

The cold air west of the Sun's place cannot rush into the heated air, on account of his approach to it ; yet that from the north and south may, and therefore which-ever side of the equator the Sun is on, the cross wind from that quarter combining with the easterly, causes it to be north-east, or south-east ; but the power of each is in proportion to their direction or current. When the Sun is on the south side of the equator, the wind blows south-east ; when on the northern, it blows north-east.

The trade winds are not regular, excepting in the open ocean; for near islands, the heat reflected from their surfaces, in some degree counteracts the natural cause of the trade winds. When this is the case, navigators call these effects land winds. They blow from shore in the day, but to shore at night; for which we can readily account, on the principles previously explained. And here ends our present scrutiny.

Oh, that I had eloquence divine, that I might express the aggregated force of the power, wisdom, benevolence, and every other attribute of the Deity, with which they strike my imagination! But, alas! that is impossible: Let us then, as Thomson elegantly and comprehensively expresses,—"*Muse his Praise.*"

CONCLUSION.

WITH what pleasure, my dear young friends, have I conducted you through the variegated field of nature! and I hope with advantage, by planting the infant shoots of that harmonious uniformity, benevolence, and order, you therein observed—and that you will cultivate the impressions you have received, so as to bring forth the fruits of those virtues, let your situation in this sublunary state be what it may, even ever so obscure: They, like the Sun's beams on the humble convolvulus, will render you conspicuous by the sweet robes of grace and harmony with which they will adorn you; or should your rank be exalted in society, by your communicable goodness, you will shine with intrinsic lustre—not with the false colouring which flattery gives to wealth, for the effects of your virtues shall proclaim your title to praise and honor. Thus, in all situations, you will find virtue to be your best friend, the most likely to make you happy in yourselves, and loved and respected by the world.

If the instruction I have been so happy in delivering to you has this effect on your minds, I shall be amply repaid for all my labour, (if I may use that term to express an occupation in which I delight, and in which I glory) in having excited your attention to those subjects, which will furnish you with arguments to confute the unbelieving—consolation to soften your sorrows—elevation of ideas to heighten your joys,—and with such a disposition of mind as will secure your happiness both here and hereafter.

END OF THE TENTH AND LAST LECTURE.

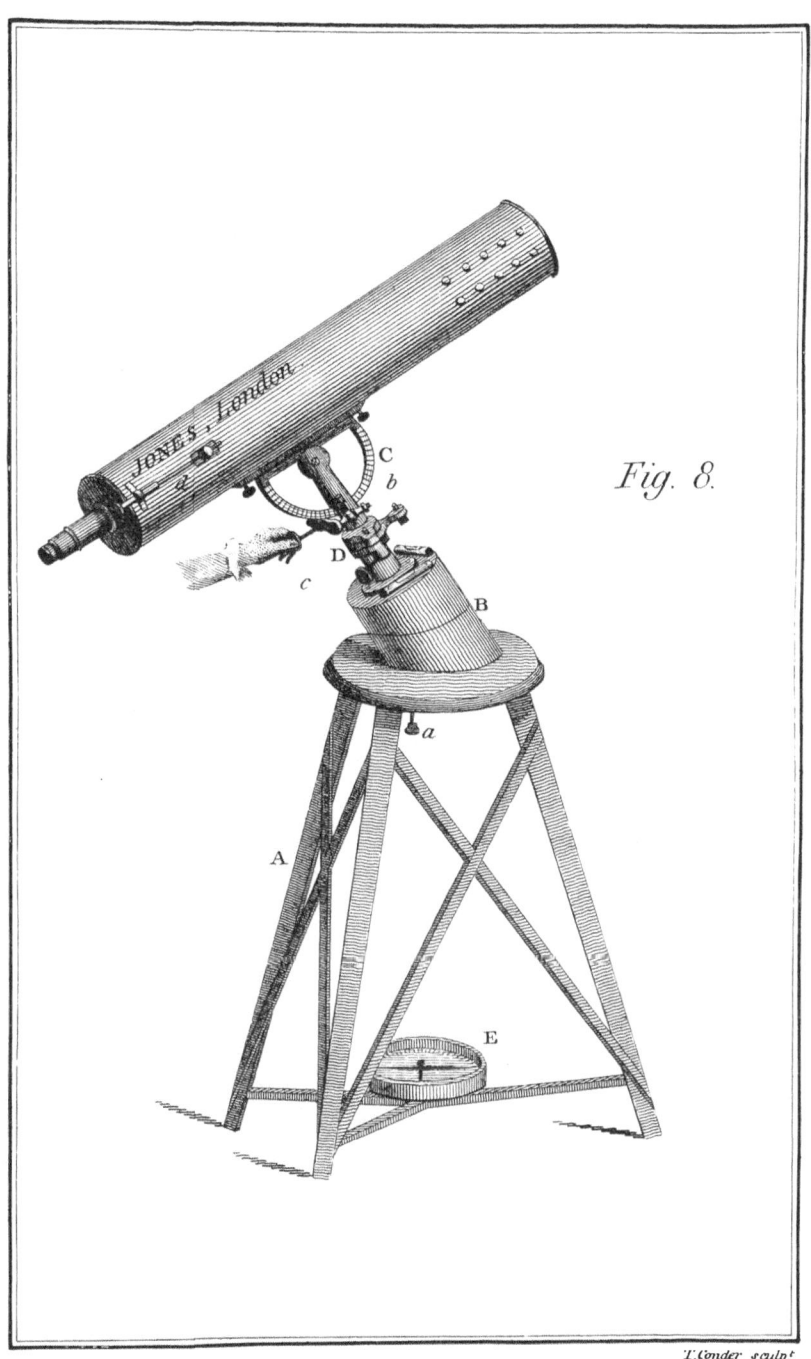

Fig. 8.

T. Conder sculp.

Plate XV.

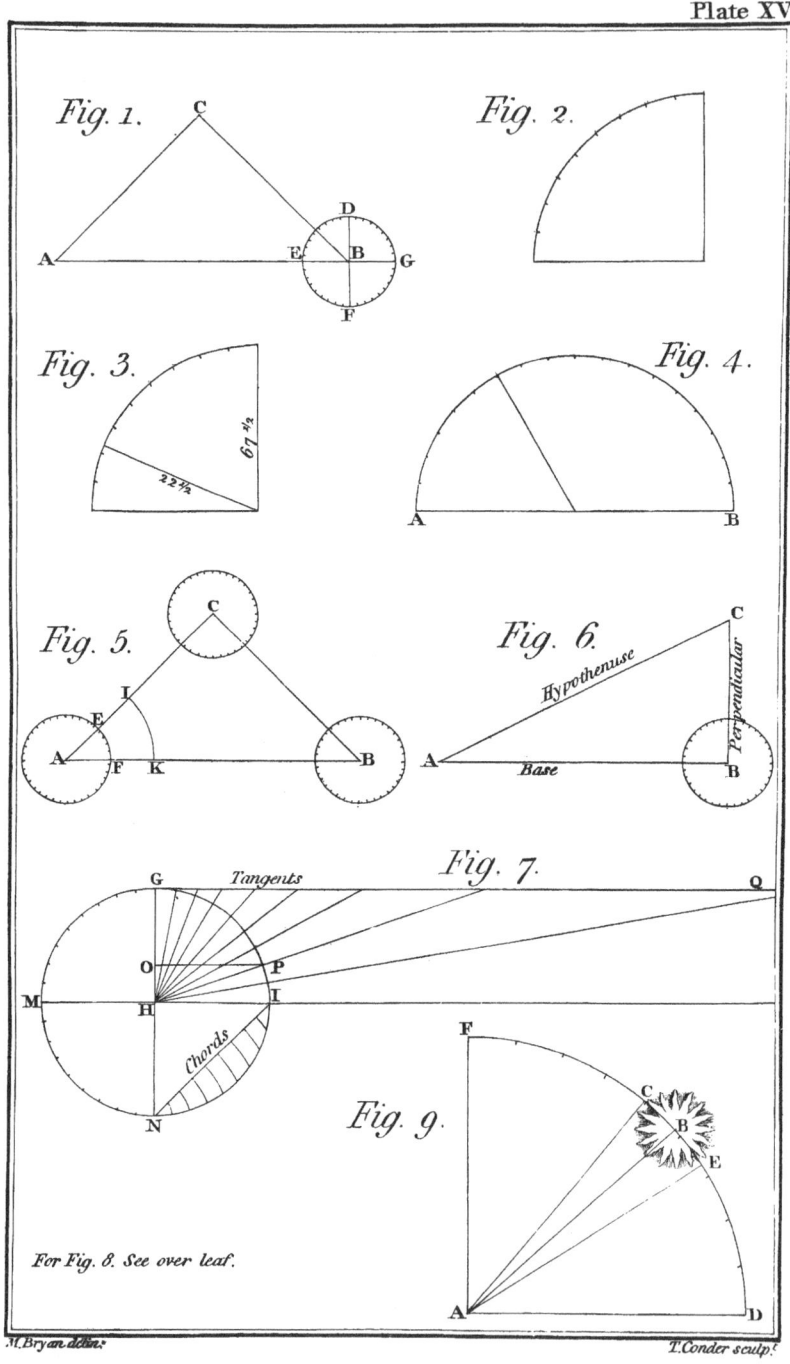

Plate XVI.

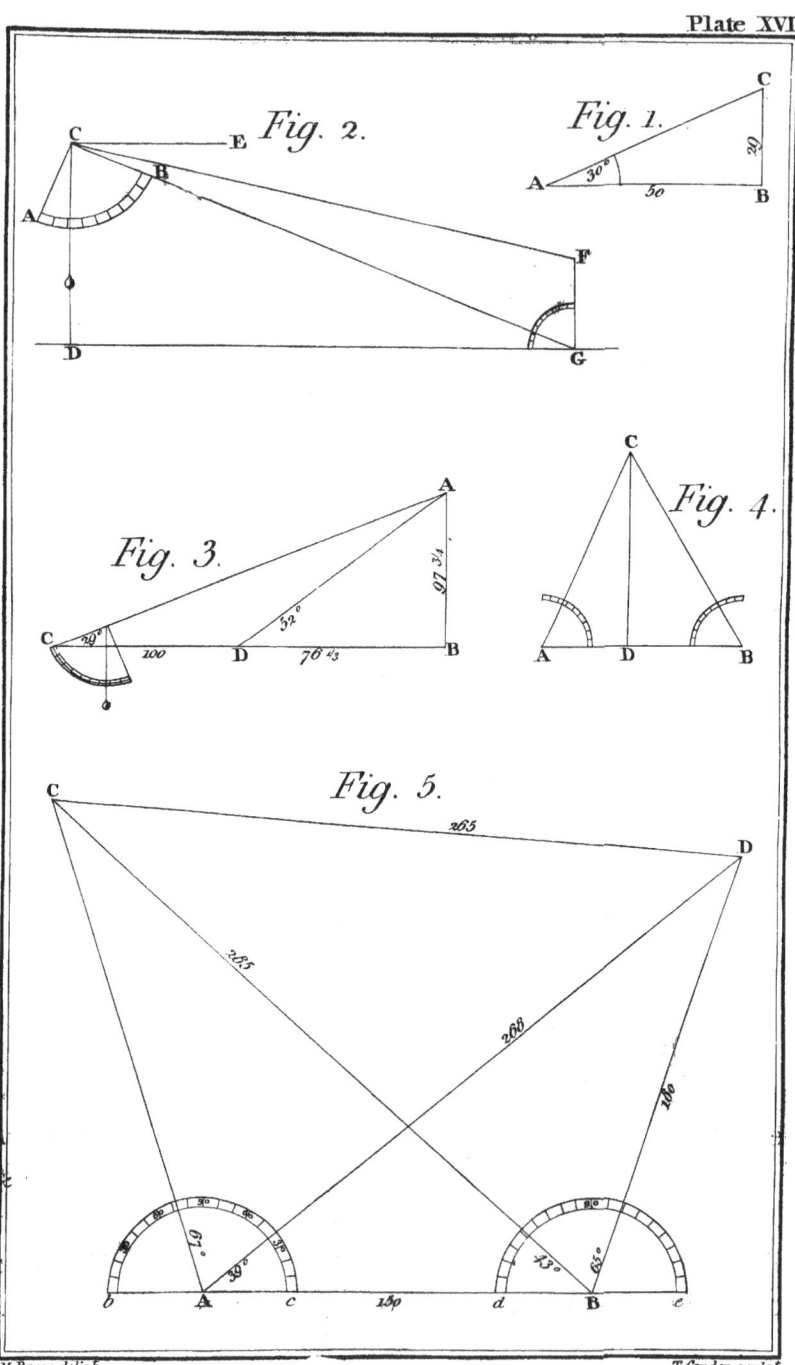

Fig. 1.

Fig. 2.

Fig. 3.

Fig. 4.

Fig. 5.

ELEMENTS of TRIGONOMETRY.

TRIGONOMETRY is that part of Mathematics which is used in estimating the sides, and ascertaining the quantity, of the angles in any triangle; a branch of knowledge absolutely necessary to Astronomers, in order to their finding the real sizes and distances of the heavenly bodies, viz. by comparing the angles they subtend with the semi-diameter of our Earth or its orbit.

Every triangle has six parts, namely, three sides and three angles, as *fig*. 1, *plate* 15, represents. A B, A C, C B, are the three sides, and A, B, C, the three angles.

The angles are always measured by the arc of a circle described on the angular point. Thus, if I want to know the quantity of the angle B, *fig*. 1, *plate* 15, I place one point of my compasses at the angular point B, and describe the arc of a circle between the two legs A B, B C; and whatever number of degrees of a circle is contained under it, is the quantity of the angle; as in this figure 45°, being the half of a quarter of the circle D E F G, the angle is said to contain 45°. Every circle being divided into 360 degrees, the number of degrees contained between two radii of a circle must be the quantity of the angle. If the number of degrees between the two legs be 90, the angle is called a right angle, and the legs are perpendicular to each other, as *fig*. 2. If the arch of the circle contained between the two legs be less than 90 degrees, it is called an acute angle, as *fig.* 3. If the arch of the circle be more than 90 degrees, as *fig.* 4, the angle is called obtuse.

An obtuse angle may contain any number of degrees between 90 and 180 degrees, or within a semi-circle; but at 180 degrees, the lines becoming straight in the same direction with each other, as at A B, the angle vanishes.

What an acute angle wants of 90°, is called its complement; as in *fig.* 3, the angle being 22° $\frac{1}{2}$, the complement is 67° $\frac{1}{2}$.

What an obtuse angle wants of 180°, is called the supplement of that angle to a semi-circle.

The three angles of any plane triangle, added together, make 180 degrees, as shewn by *fig.* 5, *plate* 15. This is the case, let the angles be drawn in any possible proportions. Hence, if one angle be right, the other two must be acute. By this rule, if two angles of any triangle be known, the third is easily found, — being only the quantity of the semi-circle, or the number of degrees wanted to make the three angles contain half a circle or 180°.

A triangle which has one right angle, is called a right angled triangle, as *fig.* 6; in which B is the right angle, and A, C, the acute angles. In right angled triangles, the longest leg is called the hypothenuse, as A C; the leg on which it rests, the base, as A B; and the other leg, the perpendicular, as C B.

When neither of the angles is a right one, the figure is called an oblique angled triangle, as *fig.* 5.

If the angles of one triangle be equal to the angles of another triangle, although the sides are of different sizes, the sides of the former will then be proportioned to the sides of the latter.

All triangles have their greatest sides opposite to their greatest angles; and if all the angles of a triangle are equal, the sides will also be equal to each other.

If any three of the six parts of a triangle be given, (excepting the three angles) all the rest may be known from them.

Any angle, notwithstanding the length of the lines which contain it, will be an equal portion of a circle; as the arc E F, which contains an equal number of degrees with the arc I K in *fig.* 5.

The place where the lines meet which form the angle, is always considered as the centre of a circle; and when letters of reference are described on an angle, the middle of the three letters should be placed on the angular point.

The diameter of a circle, or sphere, is a line passing through its centre, and meeting its circumference at two opposite points, as at G, N, *fig.* 7.

Half the diameter of a sphere, or circle, is called its radius, or semi-diameter, as G H.

The chord of an arc of a circle, or sphere, is a line drawn from the end of one radius to another, as I N; and whatever number of degrees the arc contains, the chord subtends the same number.

The sine, or right sine, is a line drawn from the extremity of an arch perpendicular to the radius drawn from the other extremity, as O P.

A tangent touches a circle only at one point, which may be any where on the surface of it, as G Q.

By the lines drawn from the centre of the circle to the tangent, or line of tangents, you perceive that no angle can be formed with it at 90 degrees, because it is parallel with the radius or semi-diameter of the figure; therefore, if a heavenly body is at so great a distance from the Earth, that it will not form an angle with the semi-diameter of it, or of its orbit, it will be impossible to ascertain its size, or its distance from it, unless there is an intermediate object near enough to it to form one. Whereas, when we can form an angle with it, or by the aid of an intermediate object, by making observation of it at two different stations on the Earth, (within the semi-diameter of the Earth or its orbit) we can discover by trigonometry its real size and distance, allowing for the visual diminution of objects at certain distances, parallax, and re-fraction.

Although a quadrant does not give a real measure of the size and the distance of the heavenly bodies, yet by ascertaining their comparative sizes and distances, Mathematicians are able, by making the proper computations necessary to the laws just mentioned, to estimate their real sizes.

Parallax is the difference of the place of a heavenly body as taken from the surface of the Earth, and supposed to be taken from its centre. Refraction making a heavenly body appear higher than its true place, and parallax making it appear lower, (as I have explained to you in my Lectures) these two effects are in general not much different from each other, and therefore little allowance need to be made for them, unless in cases of great nicety.

The angle made by a quadrant or semi-circle with the centre of a planet, by observation at two different places on the Earth, or of two different parts of its annual orbit, is used to ascertain the distance of the heavenly body from the Earth; as, by knowing what portion of the diameter or semi-diameter of the Earth or its orbit, is contained between the two places of observation, we can estimate the other parts by that quantity;—for by that rule in tri-gonometry, which fhews us when two angles and one side of a triangle are known, what the other angle and the other two sides must be, we can know the distance of the heavenly body from the Earth. And when all the angles and sides are found, such as is the proportion of those angles and sides to the one side, such will be the actual quantity in comparison of the computed, by means of the known quantity in the distances between the two places of observation—which forms the base of the triangle.

The sizes of the planets are known by an angle formed by their diameter, compared with their known distances.

The fixed stars, as I have before observed, have no sensible parallax, and therefore their size and distances are not known. This
has been settled by observations taken of them at two opposite
places of the Earth's orbit, when no sensible difference has been
perceived; the index on the plate of the quadrant always pointing
to the same degree at every place, and the plumb line always in-
tersecting the same degree on the plane of the quadrant at each,
when the sights have been directed towards a fixed star.

The forming of angles to determine the size and distance of a
heavenly body, is performed exactly as we do the size and distance
of inaccessible objects on the Earth; but the former is involved in
some computations unnecessary to the latter, namely, parallax and
refraction. Fortunately, however, for those who are not fond of
calculation, or are incapable of entering into the depth of mathematics, those circumstances have been settled on just principles, so
that we need not enter into such abstruse studies, but may find
ample gratification in the application of the knowledge we have
obtained by applying the quadrant to ascertain their observations:
Yet as even this application may not in all respects be convenient
to many, so far as relates to the apparent sizes and distances of the
heavenly bodies, on account of the journeys they must take for
such purposes, those who wish for a local use only of the quadrant,
may find sufficient amusement in observing the height of the heavenly bodies above the horizon, their situation in respect to the
constellations, their latitude and longitude, and their distances in
diameters from the Earth; which the quadrant will furnish them
with the means of, aided by trigonometry; also the size and distances of objects on the Earth, accessible and inaccessible. As to
the rest, with reference to the Ephemeris, they may satisfy themselves by the information it contains.

A telescope being a necessary instrument for the observation of
the phases of the planets, and their peculiar characteristical ap-

pearances, I recommend to all those who wish for that amusement, to purchase the sort represented *fig.* 8, *plate* 15, which is a reflecting telescope, it being I think, for the common purposes of celestial observation, the best of any within my knowledge. I have one by me, made by W. and J. Jones, Holborn, which I much prefer to a 2 ¼ feet achromatic refractor, that I formerly had. Being furnished with an equatorial or semi-circle, it answers particularly well for ascertaining the distances of bodies from each other; as those who can travel may, by the use of it, ascertain the distances of any of the planets from the Sun, or from each other, by their own observation of them; and those who do not wish to extend their speculations to those things, which have been so clearly settled as to need no further illustration, may amuse themselves by applying this instrument to terrestrial purposes, such as ascertaining the sizes and distances between objects on the Earth, either accessible or inaccessible. For common celestial purposes this instrument is particularly well adapted, having a motion regulated to the equatorial or horizontal motion of the Earth, also a vertical motion, which is nicely adjusted to shew the angle of the equator and the ecliptic. When the instrument is fixed for any particular angle of those two circles, it will preserve its parallelism through the whole of them, describing exactly the equatorial motion.

A B, *fig.* 8, *plate* 15, is a brass tube two feet in length, containing the great speculum at the end, A; and the small one near the end, B. The vertical motion is produced by the semi-circular arch C, and the pinion at D. The horizontal motion is produced by the two arms E E, and the endless screw at F. G represents a mahogany cylindrical block, cut at half the angle of the equator, 19° 14′. When one half this cylinder is turned half round, the block will give the whole circle of the equator.

The telescope, block and stand, being placed nearly in the meridian, as may be done by the compass below, allowing for the variation of it, the apparent motions of the heavenly bodies will be followed with less trouble than with a common stand; for there will be no occasion to alter the vertical directions of the instrument, in order to keep them in the field of the telescope, or by turning it either to the right or left, the equatorial motion exactly following that apparent motion in the heavens. Whereas, when using a telescope mounted in the common manner, we have both the vertical and horizontal positions of the instrument to rectify, on account of the apparent motions just mentioned.

TRIGONOMETRICAL PROBLEMS.

ALL sides or lines are estimated by a scale of equal parts; as, inches, feet, yards, miles, &c. therefore, in all problems in trigonometry, we first make a scale of equal parts, by which to estimate their length or quantity, allowing each of those parts to be equal to any quantity we please, according to the height or distance of the object we wish to make those observations of. If it is near or low, by feet or yards; if high or distant, by miles. When we want only to know the height of an object on the Earth, or its distance from it, a quadrant is perfectly adapted to the observation; but if we want to ascertain the distance between two objects that are remote from our place of observation, an equatorial will answer our purpose best.

Although I have fully explained the quadrant in my Lectures, yet it may not be unnecessary to recall attention to some circumstances of its adjustment, as probably, in applying the instrument to the purposes it is designed to answer, these might be neglected.

Previous to taking an observation of the height of a heavenly body above the horizon, and its situation in respect to the points of the compass,—render the plate of it parallel with the horizon by means of the screws, as directed; and also, make the correspondent points of the compass to answer to those in the heavens by means of a small compass box, allowing for the variation. In observing the height of a heavenly body above the horizon, direct the sights to the object, and notice the degree cut by the plumb line on the plane of the quadrant; then, as are the number of degrees from the plumb line to the end of the quadrant furthest from your eye, the same as the object is above your horizon; and as are the number of degrees from the plumb line on the plane of the quadrant to the end next your eye, so is the distance of the object from your zenith point. Also, the degree on the plate of the quadrant over which your quadrant stands, shews the point of the compass, or the meridian, on which the object is.

Thus by this instrument, at one place of observation, and by one adjustment of it, the relative height of any celestial object above the horizon of the place, such as of the Moon, planets, and stars, at night, and the Sun by day, may be found; also their distance from the zenith point, and their situation in respect to the points of the compass.

PROBLEM I.

I shall now endeavour to instruct you how to obtain the height and distance of objects on the Earth, by the use of the quadrant and the aid of mathematics.*

First, draw a scale of equal parts, or chuse one ready made. Then adjust your quadrant, and direct the sights to the top of the object, the height of which you want to know. Observe the degree of the quadrant cut by the plumb line, which will shew you how many degrees the top of the object is from the plane of your observation. If the object is accessible to you, measure its distance from the place of the quadrant, and set down what number of feet, or yards, &c. that distance contains. Next, project a triangle, by drawing a strait line horizontally for its base, or the distance of the bottom of the object from your place of observation, and from your scale of equal parts take as many for the length of

---

* In explaining the problems which concern altitudes and distances of objects on the Earth, I shall only refer to the quadrant, being of more common use, as they may be performed by it, although the semi-circle belonging to the telescope will answer the purpose as well. When we wish to ascertain the distances of two objects from each other, we must use the equatorial; and therefore I shall refer to it.

this line for the base of your triangle, as there are feet or yards in the distance measured.

Suppose the distance of your place of observation to the foot of the object be 50 yards; and you find by the quadrant that the top of the object is 30 degrees elevated above the horizon. You take from your scale 50 of the equal parts, which you apply upon the line for the length of the base of the triangle, as A B in *fig.* 1, *plate* 16. Then draw an indefinite line, B C, perpendicular to the further extremity of it. Lastly, by means of a protractor laid along the line A B, its centre at A, mark off 30° through that point, drawing the line A C, cutting the perpendicular line in C, which will be the top of the object. Then, by taking the length of the line B C between the compasses, and applying it to your scale of equal parts, it will be found to measure 29 yards nearly, which is therefore the height of the object required.

PROBLEM II.

Suppose you wish to know the height of an object which is below the level of your place of observation, as represented *fig.* 2, *pl.* 16. You must first measure the height of the tower or hill on which you stand, by observation of it, the same as of the object in the foregoing Problem; or barely by letting down a line or cord if it be a perpendicular tower, &c. Then go to the top of the hill or tower, &c. and adjusting your quadrant, fix it so that you can just see the bottom of the object, as at G, and observe the degree cut by the plumb line, which shews how many degrees it is from the end of the quadrant next your eye, that is, the depression of the object below the level of your place of observation. Then direct the line of sight to the top of the object, as at F, and again observe what depression it has from the level of your observation, by your quadrant; and the observation is completed. To construct this case, draw C D equal to the given height of the tower or hill, from your scale of equal parts; also, draw the base line D G perpendicular to D C, and C E parallel to D G; then draw the two lines C G and C F so as to form the given angles of depression; lastly, at the point G, raise the perpendicular G F, which will be the height of the object F G, as required.

PROBLEM III.

To measure the height of an inaccessible object on the Earth, you must measure a certain portion of the ground, in a straight line between that and your place of observation, in order to observe it from two stations; and by comparing the observations with the known distance measured, you will obtain both its real height and distance.

Suppose A B, *fig.* 3, *plate* 16, to be a castle at a great distance from you, the bottom of which is about level with your eye, at C. After having measured a certain space of ground in a line between it and you, take from your scale of equal parts as many as will be equal to the number of yards you have measured, and draw a line to represent that distance, containing the same number of parts of your scale. Suppose we say 100 yards for the distance measured, and that 100 parts of your scale is equal to C D. Having done so, look through the sights of your quadrant at the top of the object A B at A, and remark what degree is cut by the plumb line. Suppose 29°; then remove to D, and make the same observation; suppose 52°; then from the angular point C draw a semi-circle, and form an angle with your base line from C to A, of 29°, and from D to A 52; and where those two lines meet at A, let fall a perpendicular on the end of your base line from A to B. Then, on measuring the number of parts of your scale between D and B, you will find it contain 76⅓, which added to your measured distance of 100 feet, makes the distance of the object from your place of observation 176¼ feet; and its height will be found to be 97¼ feet.

The figure shews also, that the further the object is from us, the smaller is the angle under which it is viewed.

PROBLEM IV.

To measure only the distance of an inaccessible object, on land or at sea, you may proceed differently from the last problem, pursuing a more simple method.

Fix on any two spots on the right and left of the object, as at A and B of the object C, *fig.* 4, *plate* 16, and fix a mark at each station. Then measure the distance between the two places, and take as many parts of your scale as will answer for the number of yards, or other settled equal quantities, that are contained in such distance, and draw a line to represent that distance. Then go to one of the stations, as A, and fixing your quadrant, look through the sight to the post at the other station, and observe the degree at which the index points on the plate of the quadrant, and set it down; then turn the quadrant round ill you see the object C at the station A, observing the degree to which the index points on the plate; and set down the angle it makes with your eye at A, which is the number of degrees between the two set down; then from the extremity of your line at A, draw a line forming an angle of that number of degrees with your base line. Then proceed to your other station, as at B, and look through the sights of your quadrant at the other post at A, and observe the degree at which the index points; set it down as before, and turn the quadrant till you see the object at C, observing the difference between the degrees at which the index points in the two directions, and draw a line through that degree of it. Then where these two lines intersect each other, will be the place of the object, at C. Lastly, draw a straight perpendicular line C D from the place of the object at C, to the base line; and whatever quantity of your scale that line contains, such will be the mean distance of the object from your place of observation, which, as well as the two distances A C, B C, may be measured by your scale of equal parts.

PROBLEM V.

Although it is necessary for those who want to calculate the diameters and distances of the heavenly bodies, to remove to places of different longitude or latitude, yet those having been settled for us, we may, by the use of the quadrant, find the distance of any heavenly body from us at any time, by the number of its own semi-diameters, and multiplying them by the known number of miles they contain. As thus:—If, for instance, we wish to know how many of his diameters the Sun is distant from us at any time— with a quadrant graduated into degrees and minutes, we take the altitude of the upper and lower limb of the Sun in degrees and minutes from one station; then subtracting the one from the other, we see the difference, which shews the whole diameter of the Sun, which in this problem we will suppose to be 32'. Then by drawing a horizontal line, and on the extremity of it describing a quadrant, and drawing two lines for the two angles the Sun makes with our place of observation, we shall find the figure as represented in *fig.* 9, *plate* 15, where C E denotes the diameter of the Sun, and A C or A B his distance from the Earth. These two being measured by your scale of equal parts, and the one number divided by the other, it shews how often the Sun's diameter is contained in his distance. Or otherwise, without construction, by calculation only, thus: By proportion, as 32' is to 90 degrees, or 5400 minutes, so is C E, which suppose to be 1, to the arch of the quadrant D F,

K k

which thence comes out 169 nearly. Then, because the quadrant of a circle is in proportion to its radius, as 11 is to 7, it will be as 11 is to 7, so is 169 (equal to DF) to AB, which comes out 108 nearly; shewing that the distance in this case is about 108 times his diameter.

Having found the distance of any heavenly body in its own diameters from us, its distance in miles is found by multiplying by the known number of miles in its diameter. The Sun is supposed to be 800,000 miles in diameter, the Moon 2,175, Mercury 2,460, Venus 7,906, Mars 4,444, Jupiter 81,155, and Saturn 67,870.

In winter the Sun is about 108 or 110 of his diameters from the Earth, but in Summer the angle is less, and he must therefore be further from it.

PROBLEM VI.

To find the distance of two inaccessible objects from each other on the Earth, and from two different stations, instead of a quadrant a semi-circle is used, which is divided into 180 degrees, with a moveable index and sights, as in a quadrant, and which is added to the telescope, *fig.* 8, *plate* 15, the tube of the telescope answering in the most perfect manner for the line of sight.

Having measured the space of ground between two places, to the right and left of the two objects, as at A, B, *fig.* 5, *plate* 16, from a scale of equal parts lay down a line containing the number of parts equal to the number of yards in the ground measured, which let us suppose to be 150 yards.

Placing the semi-circle on a stand, and looking from each station to the post at the other station, and the two objects, mark the degrees at which the index points in both situations of the line of sight, and the objects at C and D. As, suppose at the station A, the sight pointing to the object D cuts off 39°, and that to the object C 108°; also at the station B, the sight directed to the object C cuts off 43°, and to the object D 108°. Then, to construct this, having made the line A B equal to 150 parts for the distance between the two places by the scale of equal parts, with the protractor laid at A and B, mark off those degrees, and through these marks draw the lines A C, A D, B C, B D ; or else, with the distance of 60 on a scale of chords, describe on the two extremities of the line at A and B a semi-circle; upon these set off the proper number of degrees taken from the same scale of chords, and draw the lines as before. The intersections of these lines will be the places of the two objects C, D. Then draw a line from one to the other, as from C to D. Lastly, on applying the compasses to all the lines, and then to the scale of equal parts, we find the distance of the two objects C D from each other to be 265 yards, from A to C 202, from A to D 268, from B to C 285, and from B to D 180.

I think I have introduced sufficient of the principles of Trigonometry for familiar purposes, and therefore I shall not extend my observations beyond these limits, by attempting to launch into the mazy labyrinths of more intricate calculation, spherical trigonometry, parallax, and refraction.

PROBLEMS on the CELESTIAL GLOBE.

THE GENERAL PRINCIPLES OF A CELESTIAL GLOBE EXPLAINED.

I SHALL subjoin such problems as are calculated to apply the observations made by Astronomers, of the different circumstances and aspects of the heavenly bodies, for different periods, by reference to the Ephemeris.

As the globes which are used to convey an idea of celestial phœnomena are differently constructed or mounted, and although I have given a description of one in my Lectures, yet I think it necessary to inform you of the general principles of a celestial globe, in which they all agree.

The celestial globe represents all the fixed stars of the ethereal regions perceived by us. These have, we know, been divided into classes, and grouped into clusters or constellations, and are distinguished by characters, according to both. There are two circles which intersect each other on this globe; one is called the ecliptic, representing the Sun's apparent path in the Heavens; and the other the equator, or the circle of our Earth, exactly in the midst between the two poles, and which, in using the celestial globe, is transferred in idea to the Heavens. The former is distinguished by eight circles drawn parallel to it on each side, including the 12 signs of the zodiac, divided into 30° each, composing together a complete circle of 360°. The latter, when transferred in idea to the Heavens, is called the equatorial, because it only appears to divide the sphere of the Heavens into two equal parts, by being equally distant from both extremities of its apparent diurnal motion; or also the equinoctial, because when the Sun is in it, the nights (and days) are equal every where. Parallel to the equatorial, and at 23 ½ degrees from it, on each side, a circle is drawn, which touches the ecliptic in two points—one on the north, and the other on the south side of the equatorial; these circles are called the tropics, or solstices. Through the two points where the tropics touch the ecliptic, a great circle of the sphere passes perpendicularly, called the solstitial colure. Where the ecliptic and equatorial intersect each other, another great circle, in like manner, passes, called the equinoctial colure. The polar circles are those drawn at 23 ½ degrees from each pole.

The horizon of the celestial globe is considered as the plane of our situation on the Earth, which is transferred in idea to the centre of this globe, because the surface of it represents that great sphere of the Heavens which appears to surround the Earth; and for which latter reason, when we wish to represent the apparent annual and diurnal motion of the Heavens, or the Sun or planets, we must turn this globe from east to west, although the real annual motion of the planets is performed in a contrary direction, from west to east.

On the plane of the horizon of this globe, (which always represents the visible horizon of any place on the Earth, when the latitude of the place is brought to touch the edge of it) are delineated the figures and signs of the zodiac; the degrees of which correspond to the days of the months, as also the months to the signs which the Sun is known to be in at each period. The principal and collateral points of the compass are also described on it. The brass circle which crosses the globe, north and south, is called the general meridian, because it serves to represent that circle which passes over every place on the Earth, when the latitude of any place on it is brought to the edge of the horizon. A dial is always a necessary appendage to a globe;—its application will be understood in the problems.

The latitude of the stars is counted on those circles of this globe drawn perpendicular to the ecliptic, and which intersect each other at the poles of the ecliptic.—Their longitude is reckoned on the ecliptic, and the lines drawn parallel to it. But the latitude of the planets is counted on that part of those lines only, which is included within the eight parallels drawn on each side of the ecliptic, because they never depart beyond those limits, north or south of it; and their longitude is counted on the ecliptic, and the eight parallels on each side of it. The longitude of the heavenly bodies is reckoned from the first degree of the sign Aries; and their latitude from the ecliptic circle, towards the poles of the ecliptic. This globe should be furnished with a quadrant of altitude, to shew the height of a celestial body above our horizon at any time.

These general principles being applicable to all descriptions of celestial globes, however mounted, will render the manner of working problems on them perfectly intelligible.

Previous to making observation, by means of a celestial globe, of the time when any celestial phœnomenon is to take place, it is necessary to know the latitude of your place of observation; which, if unknown to you, as at sea, or you have not a terrestrial globe at hand, you may discover it by the aid of the stars.

PROBLEM I.

To find the latitude of a place by the observation of two stars.

Find the Sun's place in the ecliptic for the time, either by the horizon of the globe, or else more accurately by an Ephemeris.

Bring the Sun's place on the globe to the meridian of it, which will be the situation of the Sun for 12 o'clock at noon on that day. At night, when the stars are visible, let the Sun's place at noon be at the meridian, and fix the index of your globe to 12 o'clock, that being the Sun's place for 12 o'clock on that day. Then turn the globe on from east to west till the index points to the time of your making the observation, and fix your globe there. Then observe two stars, one of which should be exactly on your meridian at the time, and the other, either at the east or west points of your horizon, as viewed in the Heavens, which may be known by a compass. Bring the star you have actually seen on your meridian to the brass meridian of your globe; then, observing carefully not to let your globe move east or west, move it, together with the brass meridian, either north or south, till the star you observed at either the east or west point of your horizon comes to that place of the horizon of your globe; and whatever elevation the pole has, when the two stars have the same situation in respect to your meridian and horizon on the globe, as in the Heavens,—such will be the latitude of the place of observation; the height of the pole above the horizon, at any place, always being equal to the latitude of it.*

---

* The longitude of places on the Earth occasions the Sun to be sooner or later upon our meridian at one place than another; so that in fixing a celestial globe to work problems by it, we only refer to the Sun's place for 12 o'clock at noon, and we know the appearance of all the celestial phœnomena will accord with the observations taken by Astronomers, if we fix the meridian of the globe to the latitude of the place of observation; as then, notwithstanding the change of longitude which causes the difference of time, the latitude being known, allows for it; or the clocks at each place being regulated by the Sun being on the meridian of each, the phœnomena will correspond with the time set down in the Ephemeris, if the latitude is brought to the horizon, the globe then representing the visible horizon of that place; and therefore the phœnomena of the Heavens will appear as set down in the Ephemeris, according to the clocks at that place, although not at the same actual time as at Greenwich, it being the same hour of the day at each place; therefore we need not regard the longitude of places on the Earth, if we have a clock or watch regulated by the Sun at the place of observation, in order to know when to look for it.

If a globe is not furnished with a compass, which is a necessary appendage, a box compass must be used, and the globe always fixed by it, allowing for the variation, and according to the longitude of the place; which being always expressed in the Ephemeris what it is at a certain degree of terrestrial longitude, which it is at Greenwich, its variation at places of different longitude may be known pretty nearly.

PROBLEM II.

To know what stars will be above our horizon for a particular evening after Sun-set.

We rectify the globe to the latitude of the place of observation, and bring the Sun's place for 12 o'clock at noon, for the given day, to the brass meridian, and fix the index to 12 o'clock; we then look in the Ephemeris for the time twilight ends. Then turning the globe westerly, agreeably to the apparent motion of the stars, we observe when the index points to the given hour; and leaving it in that situation, perceive all the stars visible at that time, as they correspond with those which are above the horizon of the globe. Again, looking for the beginning of twilight in the morning, and turning the globe till the index points to the hour twilight begins, we see all the stars which will be above our horizon during the whole of that night, and the length of time they will remain above it.

PROBLEM III.

To know what evening a particular star will be on our meridian.

We must bring that star to the meridian, and fix the index to 12 o'clock. Then turn the globe half round till the index again

L l

points to 12 o'clock, and observing what sign of the ecliptic is then at the meridian, look on the horizon of the globe for its correspondent day, which will be the Sun's place, and the day that the given star will be on our meridian at 12 o'clock at night.

PROBLEM IV.

To find the Sun's declination, which is its distance north or south of the equator.

Rectify the globe for the Sun's place, i. e. bring that degree and sign of the ecliptic the Sun is in for any day to the meridian. The degree of the brass meridian which stands over it, is his declination for that day.*

PROBLEM V.

To find the right ascension of the Sun, or a star.

When the Sun's place is at the meridian, we observe the degree of the equator which is under the meridian at the same time, as that will be the right ascension of the Sun for that day on which the Sun is in that place of the ecliptic. Then, for the observation of the right ascension of a star, we bring the star to the meridian, and the degree lying over it is its right ascension.†

---

* The globe I have is furnished with a semi-circle of declination, which is very useful, as by a sliding piece of brass on it, I can fix it to any particular spot, and see the different declination of the Sun for any hour of the day, or observe the progress of a planet more conveniently.

† The right ascension and declination of the Sun and planets change every day, but those of the stars are always the same.

PROBLEM VI.

To find the longitude and latitude of a star or planet by the Celestial Globe.

Screw the quadrant of altitude to the brass meridian, and bring it over that pole of the ecliptic which lies on the same side of the ecliptic circle on the globe with the star: Then the degree of the ecliptic circle over which the quadrant lies, is the longitude; and the number of degrees from that spot to the star, its latitude; being its distance from the ecliptic circle.

PROBLEM VII.

To find the time of the Sun's rising and setting; also his amplitude or distance from the east or west cardinal points of the horizon at those times; and the length of the day and night at any given place and time.

Elevate the pole to the latitude of the place; bring the Sun's place for that day to the brass meridian, and fix the index at 12 o'clock. Turning the globe till the Sun's place in the ecliptic touches the western edge of the horizon, observe the hour on the globe to which the index points; that will be the time of sun-setting at that latitude: And the number of degrees between the west point of the horizon and the point of it where the Sun is, will be its amplitude. Then turning the globe till the Sun's place transits the horizon on the eastern edge of it, and observing the time by the dial, that is the time of his rising; and by the distance of the Sun's place from the east point of the horizon, we know its amplitude at rising. Lastly, by doubling the time of his setting, we find the length of the day; and by doubling the time of his rising, the length of the night.

L l 2

PROBLEM VIII.

To find the beginning and end of twilight.

Rectify the globe to the latitude and Sun's place; and if the globe has a circle fixed below the horizon of it, as the one represented in the plate has, then only observe how long the place of the Sun is in passing from that to the upper side of the horizon, for the duration of the morning twilight; and the length of time it is in passing through the same space in setting, for the length of the evening twilight. But if the globe is not furnished with this wire, then fix the quadrant of altitude to the brass meridian at the zenith point, and bringing it down to the horizon, observe when the Sun's place cuts it at 18° below the horizon; this will be the beginning of twilight in the morning;—and by turning the quadrant to the other side of the meridian, you will also perceive the time of twilight ending in the evening.

PROBLEM IX.

To find the oblique ascension and descension of any star, as seen from a particular place.

Elevate the globe to the latitude of the place, and bring the star to the eastern edge of the horizon; then the degree of the equatorial cut by it, shews the star's oblique ascension or obliquity to the ecliptic. Again, turn the star to the western edge of the horizon, and the oblique descension will be found by the degree of the equator cut by the horizon.*

---

* The Sun and Planets oblique ascension and descension alter each day, to the same latitude; but the oblique ascension and descension of the fixed Stars do not alter.

PROBLEM X.

To find the rising and setting of any star, also its continuance above and below our horizon, and the time of its southing or culminating.

Rectify the globe; bring the Sun's place to the brass meridian, fixing the index to 12 o'clock; then turning the globe till the given star is at the eastern edge of the meridian, the index will shew the time of its rising; when at the meridian, it will shew the time of its southing, or passing the meridian; and when it is at the western edge of the horizon, the time of its setting; and, finally, the time of its being above or below our horizon, is found by doubling the times of its rising and setting, as mentioned in a former problem for the Sun.

PROBLEM XI.

To find the altitude and azimuth of the Sun, or of a star or planet, for a given hour.

Rectify the globe for the latitude and the Sun's place; screw the quadrant of altitude to the meridian of the globe at the zenith point, and fix the index to 12 o'clock; then turn the globe as usual, westerly, till the index points to the hour fixed upon; and laying the edge of the quadrant over the Sun, star, or planet, the degree against it will shew its altitude above the horizon for that time; and the end of the quadrant at the horizon, the azimuth of the star, or its bearing, or the direction in which it lies in respect to the north and south points of the compass at rising and setting.— The vertical circle passing through the east and west points of the compass is called the prime vertical or azimuth.*

---

* The azimuth of the Sun, or of a star or planet, means its distance from the north and south points of the horizon, at rising and setting.

PROBLEM XII.

To discover all the stars that are either rising, or setting, on your meridian, at a particular hour.

Having rectified the globe for the latitude and Sun's place, as usual, and fixed the index to the given hour; all the stars under the brass meridian are upon the meridian in the heavens, those on the eastern edge of the horizon are rising, and those on the western side of it are setting.

PROBLEM XIII.

To find the place of a planet on the globe, in order to transfer your observation of it to the heavens.

Find its longitude and latitude in the Ephemeris, for the given day; which is always set down for the noon of that day in the Ephemeris. Or else, its declination and right ascension. Mark that place with a pencil; or, if you have a semi-circle of declination, place the sliding piece of brass on it, over the place of the planet on the globe: then looking for the group of stars in the heavens it appears in, you will readily find it there. If you find by your Ephemeris that more than one planet will be above your horizon, and fit for observation on any evening, put patches on the different places, in like manner; and by directing a pointed stick from the spots on the globes to the heavens, it will point to each of them; observing previously to fix your globe, by the compass, to your meridian and latitude.

PROBLEM

PROBLEM XIV.

To find the time of the rising and setting of any of the planets, also the
time of their being on the meridian, and their bearings in respect to
the points of the compass, at rising and setting, for any evening;
having previously found their places in the Ephemeris, and
that they will be proper for observation on that evening.

Rectify the globe for the meridian by the compass, the Sun's
place, and the latitude of the place of observation, and fix the index to 12 o'clock.   Then turn the globe westerly, and observe by
the index the time when each black patch appears above the horizon on the eastern side, as that will shew its hour of rising, and
its amplitude or distance from the east point of the horizon : turning them on, you will see the time of their southing, or being on
your meridian, also of their setting, and their western amplitude.

PROBLEM XV.

To find the length of the day and night at any place between the
equator and polar circles.

Rectify the globe to the latitude and Sun's place, and fix the
index to 12 o'clock.   When the Sun's place transits the horizon on
the western side of the meridian, that will be the time of sun-
setting; when it rises above the horizon, on the eastern side, that
will be his time of rising; and by subtracting the length of the
night from 24, you will find the length of the day.   If the place
is in north latitude, you elevate the north pole above the horizon ;
if in south latitude, the south pole is raised and the north pole depressed.

PROBLEM

PROBLEM XVI.

*To find the length of the longest and shortest day at any place north of
the equator, the latitude of which does not exceed 66¼°, being
the distance between the equator and polar circles ; beyond the
latter, it would be tedious to extend our observations.*

For the longest day, rectify the globe for the latitude of the
place, also for the Sun's place when in the tropic of Cancer, (which
is the longest day at all places north of the equator) and bring it
to the brass meridian ; then fix the index to 12 o'clock, which is
the noon of that day. On turning the globe back towards the eastern side of the horizon, and observing the time that the Sun's place
touches it, you will perceive the time of sun-rising for that day, at
that latitude ; then turning the globe naturally, or westerly, till
the Sun's place touches the western edge of the horizon, the time
of sun-setting for that day will be seen, and, consequently, the
length of the day at that place.

PROBLEM XVII.

*To find the length of the longest day at any place in southern latitudes,
within 66½ degrees of the equator.*

Elevate the south pole above the horizon to the given latitude,
and bring the Sun's place when at the solstitial point, or in the
tropic of Capricorn, to the brass meridian, and work the problem
as the last. For the shortest days at northern latitudes, the Sun's
place when in the tropic of Capricorn is brought to the meridian,
and the north pole elevated. For the shortest day in southern latitudes, the Sun's place in the tropic of Cancer is brought to the
meridian, and the south pole elevated above the horizon, and the
problem is performed as for the longest days.

PROBLEM XVIII.

To find when a star rises or sets cosmically.

Rectify the globe for the latitude of the place; observe what point of the ecliptic is at the eastern side of the horizon with that star; then look on the horizon of the globe, to see what day the Sun is on that sign, which will be that in which the star will rise with the Sun, or cosmically. Then bring the same star to the western edge of the horizon, and observe what point of the ecliptic is at the horizon at the same time with it, which will shew the day that star sets with the Sun, or cosmically.

The same observations may be made of the planets, by finding their places in an Ephemeris.

PROBLEM XIX.

To find when a star or planet rises at Sun-setting, or achronically;
also, when it sets achronically, or at Sun-rising.

Rectify the globe to the latitude of the place; bring the star to the eastern edge of the horizon, and observe what point of the ecliptic is at the same time at the western edge of the horizon.— The latter being the place of sun-setting, by looking at the month and day on which the Sun is in that sign and degree, the day on which that star rises achronically will be found. If the star is brought to the western edge of the horizon, and the point of the ecliptic which is at the same time at the eastern edge of it is observed, and its correspondent day—that on which that star sets achronically, will be found.

M m

PROBLEM XX.

To find the Sun's meridian altitude for any day.

Rectify the globe for the latitude and Sun's place; and when the Sun's place is at the meridian, the degrees between that point and the horizon, counted on the brass meridian, will be its altitude for that day.

The only problem remaining to be explained to you, (being, with those I have already acquainted you, all that are most worthy your attention) is that of the Harvest Moon.

PROBLEM XXI.

Of the phœnomenon of the harvest moon.

At the time of the autumnal equinox, when the Moon is at or near the full, she rises for several nights together nearly at the same time, which circumstance being particularly useful to farmers, by affording them a longer continuance of light for gathering in their harvest with more security, it has been called the Harvest Moon.

We have every reason to suppose that this, like all other of the dispensations of Providence, was intended for our benefit, particularly so, as we know this circumstance does not occur in those situations where it would be unnecessary; and the more rationally this phœnomenon is considered, the more will the justice of the idea of its being a kind provision of Providence, appear. For is it not reasonable to suppose, that the Power, who makes the Earth team with plenty for the support of his creatures, does also take care to prevent such benefit from being lost, or defeated, which might be the case in this uncertain climate, if the produce of the fields were left too long upon the ground?

You know that the Moon, in rising, makes different angles with the horizon at different parts of her orbit. She can be full but twice a year in or near the line of her nodes, or where she makes the least angles with the Earth's orbit; which are, when her orbit intersects the ecliptic.

It is evident, when the Moon is in those signs of the ecliptic nearest the line of her nodes, she must form the least angles with our horizon; and, of course, remain the shortest time above it;— but when in those signs which make the greatest angles with it, she must remain the longest above our horizon.

The points of the ecliptic at which the Moon's orbit intersects it, are variable*. The full Moon always happens when the Moon is in the opposite sign to the Sun.

Rectify the globe for the latitude of some place in northern latitude; suppose to those on the same parallel with London, or $51\frac{1}{2}$ degrees; then bring the Sun's place for the autumnal equinox, which is the first degree of the sign Libra, to the brass meridian, and fix the index to 12 o'clock. Supposing the Sun to be in the first degree of Libra, the Moon must at the same time be at full, and in the first degree of Aries. On turning the globe westerly, till the Sun's place touches the horizon on the western edge, you will see the time of sun-setting for that day; and on looking at the eastern edge of the horizon, you will perceive the Moon rising at the same time that the Sun sets. By placing patches on the globe, agreeably to the popular mode of illustrating this phœnomenon, at every 13° of the ecliptic from Aries, to be more accurate in

* This is also a wise ordination of Providence; as, were the line of the nodes fixed to those two points of the ecliptic where it intersects the equator, we should have a total eclipse of the Moon at each autumnal equinox; or, were they fixed to any points of the ecliptic, we should twice a year be deprived of some part of the Moon's illumination, which would be inconvenient, but at no time so much as at the time of the autumnal equinox.

M m 2

working this problem, look for the place of the Moon's nodes, and place the patches according to the inclination she has at that time to the ecliptic, by placing two pins on the place of her nodes, and carrying a string round the globe in their direction; when, by placing the patches in the same at 13° from each other, you will see the time very accurately of the Moon rising; and for six days, you will find there will not be more than about an hour's difference between the time of the Sun setting and the Moon rising. This you will be able to ascertain, by allowing only one degree of the Sun's apparent motion in the ecliptic, westerly, whilst you allow 13° degrees of the Moon's apparent motion in the ecliptic, in the same direction, in 24 hours, as that will be agreeably to the true relative proportions of the ecliptic, passed over by the Earth and the Moon in that time. Then, by turning the globe on for six days, till the patch at each degree on the ecliptic for the Sun's place comes to the western edge of the horizon, and the patch at each degree of the Moon's apparent motion appears at the eastern edge of the horizon, being the time of the Sun setting and the Moon rising; by reference to the index, the difference of each, for each time, will be seen; and if the twilight is allowed for, you will perceive that there will be continual illumination for six days and nights, at the time of the autumnal equinox, in that latitude.

As the latitudes increase, the difference between the time of the Sun setting and the Moon rising, will be less; but as they decrease, the difference will be greater; as may be seen, by fixing the globe for different latitudes. And as all the times must differ according to the place of the Moon's nodes, the nearer those are to the equinoctial point, the less will be the difference of time, and the greater will be the advantage by the continuance of the phœnomena, but in respect to the latter; not so near as to cause an eclipse of the Moon.

I have an opportunity, by the use of the orrery, to amuse my pupils with many pleasing and clear problems of the aspects of the heavenly bodies, and all the phœnomena of the Heaven and Earth, with the principles of which we are acquainted; but to introduce them all in this work, would be of no utility to those who have not a similar apparatus; and to those who have, the elucidation of those circumstances may be inferred from what I have explained in my Lectures—of the planetarium, lunarium, and tellurium, which compose this orrery.

AN

EXPLANATION OF THE TABLES

CONTAINED IN

WHITE's EPHEMERIS;

ALSO OF THE TERMS USED TO EXPRESS THE CIRCUMSTANCES, OF WHICH IT IS MEANT TO GIVE PREVIOUS INFORMATION.

AN Ephemeris is a daily register of all the motions of the heavenly bodies, and their places, aspects, &c. which have been anticipated by Mathematicians and Astronomers*.

It exhibits the places of the planets, both as seen from the Sun and from the Earth; the former are called the heliocentric, and the latter the geocentric places of the planets.

---

* There is a very comprehensive Ephemeris and Nautical Almanac, printed for the Board of Longitude, under the immediate inspection of that very learned Astronomer and Mathematician, the Rev. Dr. Nevil Maskelyne, his Majesty's Astronomer Royal, which I recommend to all scientific navigators: but as it includes matter foreign to my plan, such as nautical calculations, I purpose giving a general idea of White's Ephemeris only, which contains all that is necessary to astronomical observations. The distinguishing honor the Astronomer Royal has been pleased to confer on my publication, by allowing it his respectable sanction, will ever live in my memory; and must be considered a convincing proof of the superiority and liberality of his mind, which could prompt him to encourage the faint beamy lustre of this reflected light, when likely to serve the purpose of more general information, notwithstanding the great and intrinsic illumination of his own understanding, knowing that, like the Moon, it might be of use, where the Sun beams of his intelligence had not entered.

All the calculations noted in the Ephemeris have been made according to apparent time, or 12 o'clock at noon, when the Sun is on the meridian, and to correspond to the meridian at Greenwich.

Apparent time differing from equal time, as I have explained in my Lectures; the former being that at which the Sun transits the meridian each day at noon, of the civil, or, as it is called, the natural day, which is not always in the same intervals of time; and the latter depending on the equable motion of the Earth's rotation, which is called syderial time.—To allow for the difference of those times, we are obliged to use an equatation of time; and in the right hand column of every left hand page of the Monthly Calendar, in the Ephemeris, the equatation is seen for every day in the year; that is, how much the clock is before or after the Sun.—We must add or subtract the difference, in order to find out mean time, by which to regulate our clocks. Every phœnomenon being calculated according to solar or apparent time, in the Ephemeris, it is unnecessary to equate the time for the observation of them;—all we have to do, in order to know the true time of their appearing, and being in such situations in respect to the Earth, the Sun, and each other, is to rectify for the latitude and meridian of our place. As they are all estimated for the latitude and meridian of Greenwich, which is reckoned at 51¼ degrees north latitude, and the first meridian of the Heavens, which is at the first degree of Aries; and whatever is the difference of longitude, such will be the difference of time in which the phœnomena will appear to an observer at a different longitude, allowing four minutes for every degree of longitude, because that is the difference of the Sun's time in transiting between two meridians of places which differ one degree in those longitudes, but which need not be allowed for if we have a clock properly regulated by the Sun, &c. as it is the apparent motion only which is considered in the Ephemeris, or in using a celestial globe, and therefore the longitude of places in that case are not reckoned, agreeably to the terrestrial longitude, but only according to the meridian or Sun's longitude for the time.

In the table at the bottom of each left hand page of the Monthly Calendar, the heliocentric places of the planets for every sixth day are shewn, which are their true places; Saturn's time of rising and setting is also set down under the table of equatation in these pages. In the table at the bottom of each right hand page of the Monthly Calendar, is the geocentric rising and setting of all the planets; A means afternoon, and M morning; also N north, and S south declination, which are only their apparent places and declination.

The noon of the civil day is called the first hour of the astronomical day, Astronomers beginning to count their day from the time of the Sun being on the meridian, and counting up to his again being in that situation. The upper part of the right hand pages shews the lengths of the days and nights for every sixth day, also the place of the Moon's node, and the latitude of the planets. The middle part of those pages shews the Sun's longitude or place in the ecliptic, also the Moon's longitude and latitude, together with the longitude of all the planets. D means when their motions are direct, or according to the order of the signs, and R when they appear retrograde. Every thing in the right hand monthly pages accords with the appearances only of those bodies, as viewed from the Earth, that is, their geocentric phœnomena. The upper part of the left hand pages of the Calendar, contains the Moon's age, the time of Sun rising and setting, its declination north or south of the equator, also the same of the Moon, and the equatation of time. M D in these tables means mid-day or noon.

Spec. Phœnomenorum, p. 38 and 39 Ephemeris, is intended to acquaint us with events and appearances, that we may look out for them at the proper times, and in the proper places: such as the conjunctions, oppositions, and quadratures of the Sun and Moon

N n

with the planets; the Sun's entrance into each of the 12 signs of the Zodiac; the time of the planets appearing stationary; the Moon's times of apogee and perigee; the planets aphelion, perihelion, and greatest elongations. The days of the month are in the left hand column; and on the right side of the characters are the hours, which are counted from 12 o'clock at noon of the civil day, up to 24.

Page 26 and 27, contain the times of the emersions and immersions of Jupiter's satellites; and are useful in determining the longitude of places, or their distance from the longitude of Greenwich; which application of these observations to that purpose, is fully explained in the Ephemeris; as are also every other circumstance in it, excepting those I have explained, and those which are not necessary to us, being expressive of circumstances not within my plan of instruction: Yet the terms belonging to the latter should, I think, be understood by all who use an Ephemeris;— therefore I will subjoin an explanation of their intended application in the Ephemeris.

Page 2, Chronological Notes.

Golden Number is used to express any year within the Lunar Cycle*, which is 19 years. The Lunar Cycle is that period in which the new and full Moon return in the same order, on the same sign and day, as they did 19 years before; the cause of which is explained in my tenth Lecture.

The Golden Number is put in the Ephemeris, that we may know the distance of the Moon in years from the last Cycle. To find the Golden Number for any year, we first add 1 to the year; suppose we say for the present year 1797, add 1 to it, and that will be 1798; divide that by 19, and the remainder will be 12 for the Golden Number.

---

* Cycle is a certain period or series, passing from first to last, and then returning again in the same order.

Cycle of the Sun, page 2, means a period of 28 years, which when elapsed, all the Sunday letters, and others, which are used to express the feasts, return to their former place and order; the days of the month to the same days of the week; the Sun's place in the ecliptic to the same signs and degrees, on the same months and days, so as not to differ one degree. This has nothing to do with the course of the Sun, but only with the Sundays, from which it is called Cycle of the Sun, or Sunday. To find the Cycle of the Sun for any year, as for 1797, add 9 to it, and divide by 28, and the remainder will be 14, which is the period of the Cycle for this year.

Epact, in page 2, means that settled by the Gregorian account, or Calendar, of 11 days, which means the excess of the solar year, in regard to its number of months, or the odd days there are difference between 12 solar and 12 lunar months that happen every year.

Roman Indiction, a manner of counting time among the Romans; it contains a Cycle of 15 years. At the time of the reformation of the Calendar, in the year 1582, was counted the 10th year of the Indiction. The Indiction is found by adding 3 to the year, as 1797, which makes it 1800, and dividing it by 15, when there being no remainder, the Indiction is set down in the Ephemeris 15.

Number of direction is between 35, being the limits between the earliest and latest day on which Easter can fall, which is always between March 22 and April 25. It is called the Number of Direction, because it directs when the first day of Easter happens, and how many days after the 21st of March.

The equatation of the equinoctial points, set down in p. 7, Ephemeris, is the result of a very nice calculation of the precession of the equinoxes, and nutation of the Earth's axis. All these calculations are very unnecessary for us to go into, and therefore I have only attempted to explain the nature of them.

CHARACTERS of BODIES belonging to the SOLAR SYSTEM,

AND OF

The Imaginary CONSTELLATIONS of the ZODIAC.

Char. Planets	Names of the Planets.	Char.	Twelve Signs of the Zodiac. Names.	The Letters of the Greek Alphabet, and their Names.					
☉	Sun.	♈	Aries, the Ram.	A α	Alpha	1	N ν	Nu	13
☽	Moon.	♉	Taurus, the Bull.	B β ϐ	Beta	2	Ξ ξ	Xi	14
⊕	Earth.	♊	Gemini, the Twins.	Γ γ ſ	Gamma	3	O o	Omicron	15
☿	Mercury.	♋	Cancer, the Crab.	Δ δ	Delta	4	Π π ϖ	Pi	16
♀	Venus.	♌	Leo, the Lion.	E ε	Epſilon	5	P ρ ϱ	Rho	17
♂	Mars.	♍	Virgo, the Virgin.	Z ζ ζ	Zeta	6	Σ σ ς	Sigma	18
♃	Jupiter.	♎	Libra, the Balance.	H η	Eta	7	T τ τ	Tau	19
♄	Saturn.	♏	Scorpio, the Scorpion.	Θ.Θ θ	Theta	8	Υ υ	Ypſilon	20
♅	Herſchel, or Georgium Sidus.	♐	Sagittary, the Archer.	I ι	Jota	9	Φ φ	Phi	21
☊	Aſcending Node.	♑	Capricorn, the Goat.	K κ	Kappa	10	X χ	Chi	22
☋	Deſcend. Node.	♒	Aquarius, the Water Bearer.	Λ λ	Lambda	11	Ψ ψ	Pſi	23
		♓	Piſces, the Fiſhes.	M μ	Mu	12	Ω ω	Omega	24

◄◄◄◄◄◄◄◄◄◄ ◄◄ ◀※▶ ►►►►►►►►►►►►► ►►

CHARACTERS of the AFFECTIONS of the PLANETS and their Significations;

Alſo, other CHARACTERS explained that are uſed in the EPHEMERIS.

Characters.	Names.	Signification.
☌	Conjunction.	When two planets are in the ſame ſign of the ecliptic.
☍	Oppoſition.	When two planets are in oppoſite ſigns of the ecliptic.
✳	Sextile.	When two planets are 60 degrees from each other.
□	Quartile.	When two planets are 90 degrees from each other.
△	Trine.	When two planets are 180 degrees from each other.

Mark.	Signification.	
°	A degree.	
′	A minute of a degree.	There are 60 minutes in a degree, and 60 ſeconds in a minute.
″	A ſecond of a degree.	
‴	A third of a degree.	

A. M. or m.	Anti-Meridian, or morning.
P. M. or a.	Poſt-Meridian, or afternoon.
h. m. s.	hours. minutes. seconds.

A VIEW of the SOLAR SYSTEM.

[See Plate 6.]

	Mercury.	Venus.	Earth.	Mars.	Jupiter.	Saturn.	Herschel, or Georgium Sidus.
Greatest elongation of inferior, and parallax of superior planets,	28° 20'	47° 48	* *	47° 24'	11° 51'	6° 29'	3° 4' ¼
Periodical Revolutions,	D. H. M. 87 23 15½	D. H. M. 224 16 49¼	D. H. M. 365 6 9¼	D. H. M. S. 686 23 30¾ 22	D. H. M. 4332 8 51½	D. H. M. 10761 14 36¾	D. H. 30445 18
Diurnal Rotations,	* * *	H. M. 23 22	H. M. S. 23 56 4	H. M. S. 24 39 22	H. M. 9 56	* *	* *
Inclinations of their orbits to the ecliptic,	7° 1'	3° 23' ⅓		1° 51'	1° 19' ¼	2° 30' ⅓	48° —
Place of the ascending node at present,	15° 46' ¾ of Taurus.	14° 44 of Gemini	* * *	17° 59' of Taurus.	8° 50' of Cancer.	21° 48' ¼ of Cancer.	13° 1' of Cancer.
Place of aphelion, or point furthest from the Sun,	14° 13' of Sagittarius.	9° 38' of Aquarius.	9° 15' ¼ of Capricorn.	2° 6' ¼ of Virgo.	10° 57' ½ of Libra.	45° ½ of Capricorn.	23° 23' of Pisces.
Greatest apparent diameters as seen from the Earth,	11"	58"	*	25"	46"	20"	4"

A VIEW of the SOLAR SYSTEM.

[*See Plate 6.*]

	Mercury.	Venus.	Earth.	Mars.	Jupiter.	Saturn.	Herschel, or Georgium Sidus.
Diameters in English Miles, that of the Sun being 883,217,	Miles. 3,222	Miles. 7,687	Miles. 7,964	Miles. 4,189	Miles. 89,170	Miles. 79,042	Miles. 35,109
Mean distance from the Sun in semidiameters of the Earth,	Semid. 9,210	Semid. 17,210	Semid. 23,799	Semid. 36,262	Semid. 123,778	Semid. 227,028	Semid. 453,000
Mean distance from the Sun in English Miles,	Millions. 37	Millions. 68	Millions. 95	Millions. 144	Millions. 490	Millions. 900	Millions. 1,800
Excentricities, or distance of the focus from the centre,	7,960,000	510,000	1,680,000	14,218,000	25,277,000	53,163,000	4,759,000
Proportion of heat and light, that of the Earth being reckoned at 100 degrees,	668	191	100	43	$3\frac{7}{10}$	$1\frac{1}{10}$	$\frac{275}{1000}$
Proportion of bulk, that of the Sun being 1,380,000,	$\frac{1}{15}$	$\frac{8}{9}$	1	$\frac{7}{24}$	1,424	1000	90

CATALOGUE OF THE CONSTELLATIONS

IN THE

NORTHERN HEMISPHERE.

[*See Plate* 4.]

THE Ancients distinguished only 12 Constellations, which occupy that part of the celestial concave through which the Sun appears to pass in his apparent revolution. These are called:

Aries, or the Ram, which consists of - -	46 stars
Taurus, the Bull, - - - - - - - -	109 stars
Gemini, the Twins, - - - - - -	94 stars
Cancer, the Crab, - - - - - - -	75 stars
Leo, the Lion, - - - - - - -	91 stars
Virgo, the Virgin, - - - - - -	93 stars
Libra, the Balance, - - - - - -	9 stars
Scorpio, the Scorpion, - - - - -	44 stars
Sagittarius, the Archer, - - - - -	48 stars
Capricorn, the Goat, - - - - - -	58 stars
Aquarius, the Water Bearer, - - -	93 stars
Pisces, the Fishes, - - - - - -	110 stars

The 21 Constellations, mentioned by Ptolemy, on the north-side of the ecliptic, are:

The Great Bear, which consists of - - -	105 stars
The Little Bear, - - - - - - -	12 stars
The North Pole Star, as it is called, is in the Tail of the Little Bear.	
The Dragon, - - - - - - - -	49 stars
Cepheus, a King of Ethiopia, - - - -	40 stars

Bootes, the Keeper of the Bear, - - - 53 stars
The Northern Crown, - - - - - - 11 stars
Hercules, with his Club, watching the Dragon, 92 stars
Lyra, the Harp, - - - - - - - 24 stars
Cygnus, the Swan, - - - - - - 73 stars
Cassiopeia, - - - - - - - - - 52 stars
Perseus, - - - - - - - - - 67 stars
Auriga, the Waggoner, - - - - - - 46 stars
Ophiuchus, or Serpentarius, - - - - 67 stars
The Serpent, - - - - - - - - 50 stars
Sagitta, the Arrow, - - - - - - - 13 stars
The Dolphin, - - - - - - - - 18 stars
Equus Minor, the Colt, - - - - - 12 stars
Pegasus, - - - - - - - - - 67 stars
Andromeda, - - - - - - - - - 66 stars
And the Great Triangle, - - - - - 4 stars

But several other Constellations have been added by succeeding Astronomers, of which, in the Northern Hemisphere, are:

Anser, the Goose, which consists of - - 10 stars
Aquila, the Eagle, - - - - - - 12 stars
Argo, the Ship, - - - - - - - 48 stars
Berenice's Hair, - - - - - - - 24 stars
Camelopardalus, - - - - - - - 23 stars
Cerberus, - - - - - - - - - 9 stars
Cor Caroli, Charles's Heart, is an extra Star lying between Berenice and Ursa Major.
Greyhounds, - - - - - - - - 24 stars
Leo Minor, the Little Lion, - - - - 20 stars
Lynx, - - - - - - - - - - 55 stars
Lizard, - - - - - - - - - - 12 stars

O o

Mens Menelaus, - - - - - - - -   11 stars
Musca, the Fly, - - - - - - - -   6 stars
Serpentarius, - - - - - - - - -   67 stars
Sobieski's Shield, - - - - - - - -   8 stars
Trigonus Major, the Great Triangle, - -   10 stars
Trigonus Minor, the Little Triangle, - -   5 stars

Stars of the first Magnitude, or Class, in the Constellations, are as follow :

Achernar, which is in the Constellation Eridanus.
Aldebaran, in the Head of Taurus, and which is called the Bull's Eye.
Antares, in Scorpio.
Arcturus, in the Skirts of Bootes.
Basilicus, or Cor Leonis, in Leo.
Capella, in the Left Shoulder of Auriga.
Canopus, in Argo.
Lucida Lyra, in Lyra.
Regulus, in the Heart of Leo.
Rigal, in the Left Foot of Orion.
Spica Virginis, in Virgo.
Sirius, in Canis Major.
Seven stars in the Great Bear, and 7 ditto in the Little Bear.
The Polar Star is in the Tail of the Little Bear.
Other remarkable Stars are the 7 in Taurus, called the Pleiades.

CATALOGUE OF THE CONSTELLATIONS

Noticed by the Ancients in the

SOUTHERN HEMISPHERE.

[*See Plate* 5.]

CETUS, the Whale, - - - - -   8 stars
Lepus, the Hare, - - - - - -   25 stars
Orion, - - - - - - - - -   93 stars
Eridanus, the River, - - - - -   72 stars
Canis Major, the Great Dog, - - -   29 stars
Canis Minor, the Little Dog, - - -   14 stars
Argo, the Ship, - - - - - - -   48 stars
The Hydra, - - - - - - - -   53 stars
Crater, the Cup, - - - - - -   11 stars
Corvus, the Crow, - - - - - -   8 stars
The Centaur, - - - - - - -   36 stars
Lupus, the Wolf, - - - - - -   36 stars
Ara, the Altar, - - - - - - -   9 stars
The Southern Crown, - - - - -   12 stars
The Southern Fish, - - - - - -   15 stars

To which have been added :

Anser Americanus, or Toucan, the Ameri-
can Goose, - - - - - - -   9 stars
Apis, the Bee, - - - - - - -   4 stars
Xiphias, the Sword Fish - - - -   7 stars
Charles's Oak, - - - - - - -   13 stars

Chamelion, - - - - - - - -   10 stars
Corona Meridionalis, the Southern Crown, 12 stars
Crossiers, four Stars in the form of a Cross
    near the Southern Pole.
Cyrus, the Crane, - - - - - -   14 stars
Hirundo, or Swallow, - - - - -   11 stars
Indus, the Indian, - - - - - -   12 stars
Monoceros, the Unicorn, - - - -   32 stars
Noah's Dove, - - - - - - -   10 stars
Pavo, the Peacock, - - - - - -   14 stars
Phœnix, - - - - - - - - -   13 stars
Pisces Volans, the Flying Fish, - -   7 stars
The Southern Triangle, - - - -   5 stars

NAMES of the SPOTS in the MOON.

[See Fig. 1, Plate 13.]

*	Herschel's Volcano	25	Menelaus
1	Grimaldi	26	Hermes
2	Gallileo	27	Possidonius
3	Aristarchus	28	Dionysius
4	Kepler	29	Pliny
5	Gassendi	30	Catharina Cyrillus Theophilus
6	Schikard	31	Fracastor
7	Harpalus	32	Promontorium Acutum Censorinus
8	Heraclides		
9	Lansberg	33	Messala
10	Reinhold	34	Promontorium Somnii
11	Copernicus	35	Proclus
12	Helicon	36	Cleomedes
13	Capannus	37	Snell and Furner
14	Bulliald	38	Petavius
15	Eratosthenes	39	Longrenus
16	Timocharis	40	Taruntius
17	Plato	A	Mare Humorum
18	Archimedes	B	Mare Nubium
19	Insula Sinus Medii	C	Mare Imbrium
20	Pitatus	D	Mare Nectaris
21	Tycho	E	Mare Tranquilitatis
22	Eudoxus	F	Mare Serenitatis
23	Aristotle	G	Mare Fœcunditatis
24	Manilius	H	Mare Crisium

A VOCABULARY of TERMS.

ABERRATION. An apparent change of place in the fixed stars, occasioned by the progressive motion of light,
compounded with the annual motion of the Earth in
her orbit.

ACHRONICAL, Is used to express the rising of a star
or planet at sun-set, or of its setting at sun-rise.

ALTITUDE. The height of the Sun, Moon, or stars, a-
bove the horizon of any place, reckoned on a vertical
circle from the horizon to the zenith.

AMPLITUDE, Is reckoned east and west, and is the distance of the Sun or a star from the east point of the
horizon at rising, and from the west point at setting.

ANDROMEDA. A constellation of the northern hemisphere, representing a woman chained. According
to the fable in Mythology, it is supposed to be formed in memory of the daughter of Cephas and Cassiopeia, and wife of Perseus; who being chained to a
rock, to be devoured by a sea-monster, was, at the
instigation of Juno, rescued by Perseus, whom she
afterwards married.

ANGLE. The distance contained between lines which
touch each other in one point.

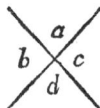 A RIGHT ANGLE, Is formed by two straight lines, one drawn perpendicular to the other.

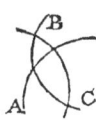 OPPOSITE VERTICAL ANGLES, Are made by two right lines which cross each other, and touch in the angular point only.

 A SPHERICAL ANGLE, Is an angle made by the meeting of two great circles, which intersect or mutually cut each other on the surface of a sphere, as the figure ABC, or that made by the ecliptic and equator.

 AN ANGLE AT THE CENTRE OF A CIRCLE, Is formed by two radii of the circle, as, *a c b*.

 A MIXED ANGLE, Is formed by a right line and a curved one.

A CURVILINEAR ANGLE, Is made by the intersection of two curved lines.

ANOMALLY, Is the angular distance of a planet from the aphelion or apogee.

ANOMALY MEAN, Is when the planet is at its mean distance from the Sun.

ANTECEDENTIA. An apparent motion of the heavenly bodies, which is contrary to the order of the signs, or from east to west.

APHELION. That place of a planet's orbit which is furthest from the Sun, called by some of the ancient astronomers the apogee, because they supposed the Earth to be in the centre of the system : In referring the planets to the Sun, we call it aphelion, using a-pogee only to express the greatest distance of the Moon from the Earth; but in reference to the Sun, we use the term aphelion, which is expressive of the same circumstance. Thus, also, perihelion means the same as perigee; only the former is used to express the least distance of a planet from the Sun, and the latter the nearest approach of the Moon to the Earth.

APOGEE. The greatest distance of a planet from the Sun.

APEX. The upper point.

APSES or APSIDES, Are the two points in the orbits of the planets, in which they are at their greatest and least distance from the Sun; and of the Moon's orbit, in which it is at its greatest and least distance from our Earth. The point at the greatest distance is called the higher apsis; and that at the nearest distance, the lower apsis. The higher apsis is also called aphelion, in reference to that place of the planets in respect to the Sun; and apogee in respect to the place of the Moon in regard to our Earth: And the lower apsis is called the perihelion, in reference to the Sun, and perigee in reference to the Earth. The diameter of the elliptical orbits, which unites the two apses, is called the line of the apses, or of the apsides.

APPULSE. Near approach to any thing.

ASPECTS, Are the respective situations of the Moon and planets with the Sun; as, opposition, trine, &c.

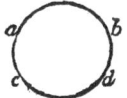 ARC OF A CIRCLE, Is any part of its circumference; as, *a b*, or *c d*.

ARCH OF A CIRCLE. A part of a circle. The same as ARC.

P p

AREA.   The surface contained within any boundary.

ARGO.   A constellation of the southern hemisphere, in
the form of a ship; which was placed there by the
Greeks, in order to celebrate the voyage of the Arganauts, a body of fifty men, of whom Jason was the
chief, who went in search, as was supposed by them,
of the golden fleece.   The fable is thus related :—
Phryxus and his sister Helle flying from the rage of
their step-mother Ino, went on board a ship, the en-
sign of which was a golden ram, and sailed in it to
Colchis, (now Mingrelia, and part of Georgia); that
Helle was drowned in the sea, which they named the
Hellespont, but which is now called the Dardanelles;
and that the Arganauts went in quest of this ship with
the golden ram, which they stiled the golden fleece :
That Jason having accomplished his design, consecra-
ted the ship to Neptune; or, according to others, to
Minerva, and that it was translated into heaven.   Sir
Isaac Newton is of opinion, that this was an embassy
sent by the Greeks during the intestine divisions of
Egypt, in the reign of Amenophis, to persuade the
nations on the coasts of the Euxine and Mediterranean seas to take the opportunity of shaking off the
yoke of Egypt, which had been laid upon them;
and that fetching the golden fleece, was only a pre-
tence to cover their design.

ARMILLARY.   Resembling a bracelet, (as the armillary
sphere) being composed of circles only.

ATMOSPHERE.   A collection of vapours surrounding the
Earth.

ATTRACTION.   The effect of an invisible natural agent,
which causes bodies to approach each other.

AURORA. The morning twilight, or the light occasioned by the refractive power of the atmosphere, which causes the light of the Sun to be seen when that luminary is about 18 degrees below our horizon.

AXIOM. A self-evident proposition.

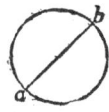

AXIS OF A SPHERE AND OTHER SOLID. A straight line passing through the centre of it to two opposite points in the circumference. It is that line round which any body will revolve, being the line passing through its centre of gravity. When this line is not spoken of in reference to motion, it is called the diameter of a sphere, as *a b*; and the centre of gravity of any other solid body, as *c d*.

AZIMUTHS. Great circles, which are supposed to pass through the zenith, and are perpendicular to the horizon.

BAROMETER. An instrument constructed for ascertaining the weight of the atmosphere, and its variation, by the rising and falling of quicksilver, ·or other fluid, in a tube. When the air is heavy, as in clear weather, the quicksilver rises in the tube by the pressure of the air on the quicksilver in the reservoir: When the air is light, or foul, the quicksilver falls in the tube, shewing the approach of foul weather.

CARDINAL POINTS. The east, west, north, and south points of the compass. Of the ecliptic, the first degrees of aries, cancer, libra, and capricorn.

CAMELEOPARDALUS. A particular species of animal peculiar to Africa. A new constellation of the northern hemisphere.

CENTRIPETAL FORCE. That by which a revolving body endeavours to approach the centre of its orbit.

CENTRIFUGAL FORCE. That by which a revolving body endeavours to fly off from the centre of its orbit.

CENTAUR. A southern constellation. In fable, Centaurs were supposed monsters, the upper part human and the lower like a horse. They always carried clubs, and were skilful in the use of the bow. The fabulists tell us, that Hercules vanquished these monsters.

CEPHEUS. A king of Æthiopia, father of Andromeda. According to the fable, he was taken up into heaven along with his wife Cassiopeia and his daughter, and placed near the Little Bear.

CERBERUS. The name given by the ancients to express a dog with three heads and three mouths, which they supposed to be stationed at the gate of the infernal regions; and that when Hercules went down to bring Alcesta back to her husband, from whom she had been taken, he bound this dog with a chain, and brought him into the upper regions. This forms a small constellation of the northern hemisphere.

CHORD. A chord is a right line which extends between the two extremities of the arc of a circle.

CHAOS. Confusion, irregular mixture.

CIRCLE. A circle is a plane figure included within one line, called the circumference, which is in every part of it equidistant from the centre.

CLOUD. A collection of vapours in the air.

COLLATERAL. The collateral points in cosmography are the intermediate ones between the cardinal.

COMA BERENICE. A modern constellation of the northern hemisphere, lying between the tail of the Lion and Bootes. It is said to have been formed by the astronomer Conan, in order to console the queen of Ptolemy Evergetes for the loss of a lock of her hair, which had been stolen from the temple of Venus, where she had deposited it, having dedicated it to that goddess on account of a victory obtained by her husband.

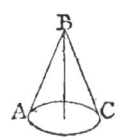

CONE. A geometrical solid figure, consisting of straight lines that arise from a circular base, and gradually declining from the surface and circumference of the base, meet in a point which is directly over the centre of the base, like the figure A B C.

CONSEQUENTIA. A motion from west to east, or according to the order of the signs.

CONTINENT. A main land which is not interrupted by seas; and so called, in contradistinction to island and peninsula.*

CONVERGING. Tending to one point from different places.

COSMICAL. A term used to express the rising or setting of a planet or star with the Sun.

COSMOGRAPHY. The science of the general affections or system of the world.

CUBE. A number in arithmetic arising from the multiplication of a square number by its root. As thus: Suppose any number, as 2, be multiplied by itself, the product will be 4, which makes the square. Then

* An Island is a tract of land surrounded by water; and a Peninsula is a portion of Land joining to a Continent or main Land by a narrow neck, the rest being encompassed by water.

if that square number be multiplied by its root 2, that will make 8 ; which latter is called the cube or third power of 2 ; and with respect to the 8, the number 2 is called the cube root.

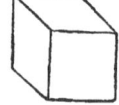

CUBE, In geometry. A regular body consisting of six square and equal sides, and the angles of course all equal and right.

CULMINATING. When the Sun or a star is passing the meridian of any place.

CYCLOID. A certain curve formed by the motion of a wheel. If a pendulum be made to vibrate in this figure, its oscillations will be all performed in equal times.

CYLINDER. A round column which is neither swelling nor diminishing.

DAY, NATURAL OR CIVIL. The time in which the Earth completes its revolution on its axis.

DAY, ARTIFICIAL OR SOLAR. The time the Sun is above the horizon of any place.

DAY, ASTRONOMICAL. That computation of time which begins when the Sun transits the meridian of any place, and ends at his return to it again.

DECLINATION. The distance of a star from the equator, north and south. The least declination of a star is its shortest distance from the equator.

DEGREE. The 360th part of a circle. A degree of the latitude of the Earth contains 69 English miles, or 60 geographical or Italian miles. The quantity of a degree of longitude differs at every degree of latitude, from the equator to the poles. At the equator, a degree of longitude is supposed to contain 69 English miles; from whence, the quantity of a degree of longitude continually diminishes at every degree of latitude, till it is nothing at the poles. In Dr. Hutton's Dictionary, vol. 1, p. 364, there are tables for each degree of latitude.

DENSE. Close, approaching to solidity.

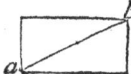 DIAGONAL. A right line which unites two opposite angles, as *a b* in the figure.

DIAMETER. The line which passes from one side of a circumference, through the centre, to the other side.

DIGIT. A measure by which the part of a luminary eclipsed is ascertained. The body eclipsed being always supposed to be divided into 12 parts called digits, as many of those parts as are eclipsed, so many digits is the body said to be eclipsed.

DIRECT, In optics, means the immediate effect or direction of the primary rays; in contradistinction to those of reflected or refracted ones.

DISC. An apparently plane round surface.

DIVERGENCE. The spreading out of light as it recedes from a centre.

DIURNAL. Relating to a day, or 24 hours.

DOMINICAL OR SUNDAY LETTER. One of the first seven letters of the alphabet; the application of which is fully explained in the Lectures.

ECLIPTIC CIRCLE. The Sun's apparent path in the heavens. It is called ecliptic, because eclipses happen when the Moon is in this circle.

EFFICIENT. Causing effects.

EFFLUVIA.   Small particles flying off from bodies.

ELONGATION.   The apparent distance of a planet from the Sun, as seen from the Earth.

EMERGING.   Rising from obscurity.

EMERSION.   The re-appearance of a planet or the Sun, after it has been eclipsed.

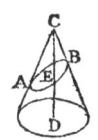

ELLIPSIS OR ELLIPSE.   An oval figure formed by cutting a cone obliquely to its axis.   C D is a cone, and A B an ellipse cut obliquely to the axis C D of the cone.   The more obliquely the cone is cut to the axis, the longer the ellipse will be.

EPACT.   The Moon's age at the beginning of the year, or the difference between the course of the Moon and the Sun; as explained in the Key to the Ephemeris.

EQUATOR.   The circle which equates or divides the globe of our Earth into two equal hemispheres, being equally distant from the north and south poles.   When we speak of this circle, in reference to the Sun or stars, we call it the equinoctial.

EQUINOCTIAL POINTS OR EQUINOXES.   Those two points of the globe where the ecliptic and equatorial intersect each other.

EQUILIBRIUM.   Equality of weight or power.

EQUI,   Means equal, and is compounded with other words, to denote equation or equality.

ETHER.   An element considered to be more subtile than air or light, and an intermediate agent which is supposed to fill the regions beyond the atmosphere of the planets.

FLUX.   Flowing in.

Focus. Where rays meet.

Focii. The plural of focus.

Galaxy. The milky way; an innumerable quantity of fixed stars.

Geometry. The science of extension, quantity, or magnitude.

Geometrical. Explained or described on the principles of geometry.

Graduated. Marked with degrees.

Gravity. Weight, attraction, tendency to a centre.

Gravity, particular, Is that whereby bodies descend towards the centre of the Earth.

Gravity, universal, Is the existence of the same principle in the heavenly regions.

Heliacal. The heliacal rising of a star, is its transiting the horizon just before the Sun in the morning. Helical setting of a star, is setting with the Sun, or when it is hid by the Sun's beams on an evening, so as not to be seen after that luminary has sunk below the horizon.

Heliocentric. The place of a planet, in which it would appear from the centre of the Sun.

Hemisphere. The exact half of a sphere.

Horizon. The line that terminates our view of the heavens. The rational and the sensible horizon are the same in relation to the fixed stars, on account of their immense distance from us: but in regard to the planets, they are not; which produces what is called the horizontal parallax of the Sun and planets, when we are considering them.

Q q

HORIZONTAL LINE. That drawn parallel to the horizon.

HORIZONTAL PARALLAX. The difference between the real place of a planet or the Sun, as viewed from the centre of our Earth; and its apparent place as viewed from its surface, when the luminary is at the horizon. So that in regard to the true place of a planet, as viewed from the surface of the Earth, we allow a parallax or this difference; and when we take the distance of the Sun or a planet from it, this parallax shews that distance.

HYDRA. In Mythology, a terrible monster which was destroyed by Hercules. The poets and painters describe it sometimes as a serpent with many heads, and sometimes a human head with serpents twined about it instead of hair. This fable is supposed to have originated in a great quantity of serpents which infested the marshes of Lerna, near Mycene, which Hercules destroyed by burning the seeds in which they were ambushed. It is a southern constellation which is intended to represent this imaginary monster.

HYPOTHESIS. A supposition.

IMMERSION. The entrance of the Earth or Moon into each other's shadow.

INCLINATION. The angle made by the orbit of one planet with that of another.

INCIDENCE. The striking of one body on another. The angle made by the direction of the moving body with the other, is called the angle of incidence.

INCORPOREAL. Without body.

INERT. Motionless in itself.

INGRESS. The Sun's entrance into any sign or degree of the ecliptic.

INTERCALLARY. Insertion of something supernumerary, as the intercallary day in leap year.

INVERSELY. Contrary ways, or in a contrary order.

LATENT. Secret, concealed.

LATITUDE, Of places on the Earth, is counted north and south from the equator. The latitude of a star, is its distance from the ecliptic, north and south, towards the poles of the ecliptic. The latitude of a planet, is its distance from the ecliptic, north and south, within certain boundaries.

LENS. A glass; plane, convex, or concave.

LIBRATION. Trepidation; a term used to express the irregular motion of the Moon's axis.

LONGITUDE, Of places on the Earth, in England, is counted from the meridian of London or Greenwich, east and west, upon the equator. But longitude of the planets and stars, is counted on the ecliptic, east and west from the intersection of the ecliptic and equator, from Aries or the equinoctial point.

LUCID. Bright or glittering.

LUNAR ASPECTS, Are those of the Moon with the Sun and planets; as, opposition, trine, quartile, &c.

LIZARD. A creeping small animal; a celestial constellation.

LYNX. A sharp-sighted spotted beast; a constellation.

MATHEMATICS. That science which contemplates whatever is capable of being numbered or measured.

MATHEMATICIAN. One versed in the mathematics.

MEAN DISTANCE. That between the greatest and least distance of a planet from the Sun.

MEAN MOTION. That between the swiftest and slowest motion of a heavenly body, and which would take place if they moved in a perfect circle.

MERIDIAN. That great circle of the sphere passing over any place, and through the north and south poles of the world, being perpendicular to the equator. This is the place of the Sun at mid-day or noon at every place.

MERIDIAN, FIRST. The first meridian is arbitrary ; most astronomers fix it at the capital city or place of observation to which they belong.

MERIDIAN, MAGNETICAL. That to which the poles of the magnet point.

METAPHOR. A figure of speech ; a simile comprised in a word.

METAPHORICAL. Figurative ; not real.

METAPHYSICIAN. One who considers the general affections of incorporeal things.

METEORS. Appearances in the sky of a transitory nature, such as clouds, thunder, &c.

MIRROR. Any thing which represents the images of objects by reflection.

MODIFICATION. Change in the form, mixture, &c. of any thing.

MOMENTUM. The quantity of force or motion in a moving body.

MINUTE. In time, the 60th part of an hour.

MINUTE. In motion, the 60th part of a degree.

MONTH, LUNAR, Is also called a periodical month, being the time in which the Moon revolves round the centre of gravity between herself and the Earth. She performs this revolution in 27 days, 7 hours, and 43 minutes, which is therefore the length of the lunar month.

MONTH, SYNODICAL, Is the period in which the Sun, Moon, and Earth are again in a direct line with each other. It consists of 29 days and a half, which is the period of each new Moon.

MONTH, SOLAR AND CALENDAR. The time of the Earth's moving through one sign or 30 degrees of the zodiac, which causes the apparent motion of the Sun through that space, and is performed in thirty days and a half, at a mean computation.

NADIR. The point immediately under our feet, opposite to the zenith.

NEBULA. A mist or fog, or cloud-like appearance.

NEBULOUS STARS. Those perceived only by a faint light, by which they are not distinctly seen without the aid of glasses; such as the milky way.

NOCTURNAL. Nightly.

NODES. The two points of the ecliptic intersected by the orbit of a planet.

NUCLEUS. The central part of a comet.

OBLATE. Flatted at two opposite points of a spheroid, called the poles.

OBLIQUE. Not directly, but slanting.

OCCULT. Hidden.

OCCULTATION. Obscuration.

OCTANT. A term used for that situation or aspect of two planets when they are 45° from each other.

OMNISCIENT. Infinitely wise, knowing all things.

OPAKE. Not transparent.

OPPOSITION. That situation of two heavenly bodies in respect to each other when they are 180° distant, or in opposite signs of the ecliptic.

ORB. A circular figure.

ORBIT. The line described by the revolution of a planet, or the curve in which it moves.

ORGANIZATION. Construction, in which the parts are so disposed as to be subservient to each other.

ORIENTAL. Eastern.

ORION. A southern constellation consisting of 93 stars. In fable, a giant and famous hunter, who having told Diana that he could kill more beasts than she could, she raised a scorpion which bit him, so that he died of the wound. But Jupiter translated him into the heavens, and made that constellation of him which goes by his name.

OSCILLATION. Moving backwards and forwards like a pendulum.

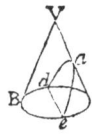

PARABOLA. A curved figure arising from the section of a cone when cut by a plane parallel to one of its sides, as the section *a d e*, parallel to the side V B of the cone.

PARALLAX. The difference of place of a planet, as viewed from the centre of the Earth or Sun, and as viewed from their surfaces.

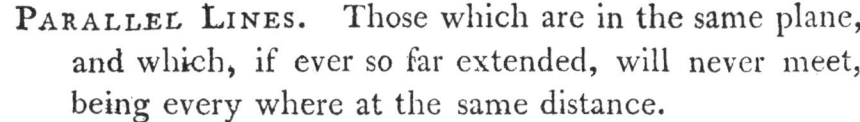

PARALLEL LINES. Those which are in the same plane, and which, if ever so far extended, will never meet, being every where at the same distance.

PARALLELOGRAM. A quadrangle, the opposite sides of which are parallel.

PARALLELS, Of latitude, are the lines drawn parallel to the equator of a terrestrial globe, and supposed to be drawn parallel to the ecliptic of a celestial sphere.

PENUMBRA. An imperfect shadow, or a partial one, which is always perceived between the perfect shadow and full light, in an eclipse; also between the perfect shadow and light of any illuminated object which casts a shadow behind it.

PERIGEON. That situation of a planet in its orbit nearest to the Earth.

PERIHELION. That point of a planet's orbit in which it is at its least distance from the Sun.

PERIPHERY. The circumference of any regular figure, as a circle.

PERSEUS. The son of Jupiter. He obtained the helmet of Pluto, the buckler of Minerva, and a faulchion of Mercury, with wings for his feet. He rescued Andromeda from the monster which was sent abroad by Neptune, and by whom she was to have been devoured: She was chained to a rock by the command of Cepheus, her father, for that purpose, in order to stop the ravages this monster had committed; but just as he was going to seize the destined prey, Perseus, mounted on his horse Pegasus, flew the monster and delivered the lady.

Periodic.   Happening at stated times.

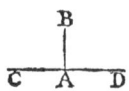

Perpendicular.   A line drawn in a direction exactly transverse, or at right angles, to another line, as A B, which is perpendicular to C D.

Phases.   The different forms of the illuminated part of a planet as seen from the Earth.

Phasis.   The appearance exhibited by any body in its changes, as that of the Moon.

Philosophy.   Knowledge, natural and moral; the course of the Sciences.

Phlogiston.   The inflammable part of a body.

Phœnix.   One of the constellations.   A most rare bird, and the only one of its species.   The fable runs thus: That after 1000 years, a fire proceeds from its nest, and consumes it; but from the ashes an egg is generated, from which the bird is again renewed. — Or thus: That after 1000 years, it plucks off its own feathers, and dies by mere decay of nature; but that it first builds itself a funeral pile of the branches of sweet-scented trees, upon which it sits down and expires: That from its ashes a worm proceeds, which changes into a bird; and, according to Heroditus, it takes up the remaining ashes of its parent, and embalms them in a mass of myrrh.   This bird is described as very beautiful and rare, and about the size of an eagle; the feathers of its head and neck to be of a gold colour, and its tail purple and carnation.

Phœnomenon,   (In the plural, Phœnomena)   Extraordinary appearance in nature.

Phosphorus.   A chemical substance which, when exposed to the air, takes fire, or shines in the dark.

PHYSICAL. Relating to matter, or natural philosophy; substance, in opposition to spirit.

PLANE. A flat superfice, neither convex nor concave.

PLANETS. Those celestial bodies in our system which have an evident motion peculiar to themselves : They are supposed to be habitable worlds, like our Earth, and to move round the same Sun, their vivifying principle.

PLANO-CONVEX, OR PLANO-CONCAVE. Flat on one side and a portion of a sphere on the other.

POLAR. Lying near the poles.

POLARITY. Tendency to the poles.

POSTULATUM. Position assumed without proof.

PRIMARY. Original, first.

PRISM, Is a figure of any number of sides, and two opposite equal bases or ends. That used for optical purposes consists of only three sides. It separates the rays of light which come blended from the Sun, by the different degrees of its refractive power, occasioned by its form, by which a ray of light received on it, passes more or less obliquely through it ; and the momentum of their individual motion arising from the different sizes of the particles, causes some to be turned more aside than others, by an interposed substance.

PROBLEM. A question proposed to be done.

PROJECTILE. A body put in motion by one force only, which always causes it to proceed in a straight line.

PROMONTORY. A high land jutting into the sea.

R r

QUADRANT. The fourth part of a circle. An astronomical instrument.

QUADRANGLE, or QUADRILATERAL. Every plane figure which is bounded by four right lines is called quadrilateral, or a quadrangle.

QUADRATURE. The first or last quarter of the Moon.

QUADRUPLE. Fourfold.

QUARTILE. When the planets are 90 degrees or a quarter of a circle from each other.

QUIESCENCE. A state of rest.

RADIANT. Shining.

RADIATION. Emission of rays.

RADII. In the form of rays. The plural of radius.

RAREFACTION. Lightness or thinness, procured by the extension of a body, by which it is made to occupy more room.

RATIO. Proportion.

RATIONALE. A solution or account of the principles of some opinions, action, hypothesis, phœnomenon, &c.

RAY OF LIGHT. The least particle of light that can be separately impelled.

RAYS, CONVERGING. Those which tend to or meet in a point. All convex lenses make the rays converge, and concave ones make them diverge.

REACTION. The resistance any body makes to the force impressed on it. The action and re-action of bodies upon each other are equal.

RECEIVER OF AN AIR PUMP. A glass, under which all the air contained in any thing may be extracted by the operation of the pump.

RECTILINEAL. Right lined.

REFLECTION. The act of throwing back.

REFLUX. Flowing out.

REFRACTION. The breaking or bending out of the straight course in a moving body. The variation of a ray of light from the straight line which it would have passed on in, had not the rays passed through a new medium.

REFRIGERATION. The property of cooling.

REGION. Place of any thing.

REGRESS. Passing backwards.

REGULATOR. That part of a machine which limits or regulates its motion.

RESPIRATION. The act of breathing.

RETINA. The last or innermost coat of the eye, being that on which the images of objects are painted: The crystaline humour collects all the rays from any luminous object, and throws them on the retina, which is placed at its focus.

RETROGRADE. Contrary to a direct motion.

RETROGRESSION. Moving backwards.

ROOT. A number multiplied by itself. See CUBE.

ROTARY. Whirling round.

REVERBERATION. The act of beating back.

REVOLUTION. The circular course of any thing which returns to the point from whence it began to move.

SATELLITE. A small planet which revolves round a larger.

SCIENCE.   Knowledge founded on demonstration.

SECOND.   The 60th part of a minute in time, and the same in motion.

SECTION.   A cut or division which separates one part from the rest.

SECTOR.   A mathematical instrument used for laying down or measuring angles.

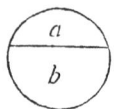

SEGMENT OF A CIRCLE.   The part contained under the arch, and a chord of that arch; as, *a* or *b*.

SEMICIRCLE.   Half a circle, as **A B.**

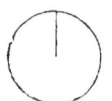

SEMIDIAMETER.   Half the diameter of a circle.   It is sometimes called the radius or whole sine.

SENSATION.   Perception by the senses.

SENSIBLE.   Perceptible by one of the five senses.

SENSITIVE.   Perception without reflection.

SENSORIUM.   Seat of the senses, or where their impressions are digested.

SEXTILE.   The aspects of the planets when they are 60 degrees from each other.

SOLAR.   Belonging to the Sun.

SOLID.   A solid has all the geometrical dimensions, viz. length, breadth, and thickness.

SOLSTITIAL POINTS.   The Sun's most northern and southern limits in the ecliptic.

SOURCE.   Spring of action.

SPECIFIC GRAVITY.   Comparitive weight; bulk for bulk.

SPECIES.   Sort.

SPHEROID. A body oblate or oblong, approaching to the form of a sphere.

SPIRAL. Winding, curved.

SQUARE, Is a rectangle having all its sides equal, being composed of four right angles and four equal sides.

SQUARE NUMBER. The product of a number multiplied by itself, as 4 multiplied by itself, or 4 times 4 are 16, which is the square of 4; and 4 is the square root of 16.

STRATA. Beds, layers.

STELLATED. In the form of stars.

SUBDIVIDED. A part divided into more parts.

SUBLUNARY. Beneath the Moon.

SUBTEND. To extend, under certain limitations.

SUBTERRANEOUS. Laying under the surface of the Earth.

SUBTILE. Thin, fine, rare.

SUPERFICE. The surface of a solid. Lines are the boundaries of a superfice. A point is the extremity of a line. A plane is a level or flat superfice.

SYDEREAL. Something relating to the stars; as, sydereal day, sydereal year, &c.

SYMBOL. An emblem of some circumstance which is expressed under a familiar sign or figure.

SYNOD. A conjunction of the heavenly bodies.

SYNONIMOUS. The same thing expressed in different ways.

SYSTEM. A scheme which unites many things in order.

SYSTEMATIC. In a regular way.

SYZYGIES. The points of the Moon's orbit in which she is at new and full.

TACITLY. Silently.

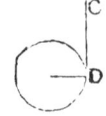

TANGENT OF A CIRCLE, Is a straight line drawn without the circle perpendicular to some radius, and it touches the circle but in one point; as, C D.

TANGIBLE. Perceptible by the touch.

TELESCOPE. An instrument furnished with lenses or mirrors, so placed as to bring the images of distant objects near to the eye.

TEMPERATE. Moderate.

TEMPERATURE. Degree of heat.

TERRAQUEOUS. Composed of land and water.

THERMOMETER. An instrument which shews the degree of heat, or cold, by the expansion or contraction of a certain quantity of quicksilver.

THEORY. Not practice.

TORRID. Burning. The zone of the Earth between the tropics, where the rays of the Sun fall perpendicularly.

TRANSIT. The passage of a planet over the face of the Sun or another planet, or some line of the heavens.

TRANSVERSE. In a cross direction.

TRIANGLE. A triangle has three sides and three angles. For a full definition of a triangle, and the different sorts of them, see Elements of Trigonometry.

TRIGONOMETRY. The art of measuring and estimating triangles.

TRINE. An aspect of the planets in which they are 120 degrees, or the third of a circle, from each other.

VACUUM. A space devoid of air or matter. It is supposed there is no such thing in nature as a perfect vacuum; that called so, and which is procured by the air pump, is only so in respect to the atmosphere of the Earth, not to the etherial fluid, which is supposed to pervade all things: Nor is it in this instance even so in regard to the matter of light, as that fluid must also exist under the receiver when the air is exhausted.

VAPOUR. Any effluvia from terrestrial substances that mingles with the air.

VIBRATE. To move to and fro; to tremble.

VERTICAL. Perpendicular to the horizon.

VESPER. The evening.

VITAL. Essential to life.

VOLCANO. A burning mountain.

UMBRA. A perfect shadow. Penumbra means an imperfect one.

UNDULATION. A waving motion.

ZENITH. The point directly over our heads.

ZODIAC. A girdle which surrounds the apparent sphere of the heavens, and extends 9 degrees on each side of the ecliptic, including the orbits of the planets.

ZONE. A girdle or belt.

F I N I S.